INTERIOR PLANTSCAPING

Principles and Practices

Join us on the web at

www.cengage.com/community/agriculture

INTERIOR PLANTSCAPING

Principles and Practices

James M. DelPrince, PhD, AIFD, PFCI

Professor

Mississippi State University

Australia • Brazil • Japan • Korea • Mexico • Singapore • Spain • United Kingdom • United States

Interior Plantscaping: Principles and Practices
James M. DelPrince

Vice President, Careers & Computing: Dave Garza

Senior Acquisitions Editor: Sherry Dickinson

Director of Development, Careers: Marah Bellegarde

Senior Product Manager: Juliet Steiner

Editorial Assistant: Scott Royael

Vice President, Marketing: Jennifer Baker

Marketing Director: Deborah Yarnell

Marketing Manager: Matthew Williams

Senior Production Director: Wendy A. Troeger

Production Manager: Mark Bernard

Senior Content Project Manager: Elizabeth C. Hough

Senior Art Director: David Arsenault

Cover Image: "© cienpies/Fotolia"

Interior: Title Page and header art for: Table of Contents, Preface, About the Author, Acknowledgements, Dedication, Glossary, Appendix, and Index: © iStockphoto/Marian Pentek. (This image is *Pilea* sp. Friendship Plant.) Green Tip: © Ljupco Smokovski/www. Shutterstock.com

For product information and technology assistance, contact us at
Cengage Learning Customer & Sales Support, 1-800-354-9706
For permission to use material from this text or product,
submit all requests online at **www.cengage.com/permissions.**
Further permissions questions can be emailed to
permissionrequest@cengage.com

Library of Congress Control Number: 2012931964

ISBN-13: 978-1-4354-3963-4

ISBN-10: 1-4354-3963-5

Delmar
5 Maxwell Drive
Clifton Park, NY 12065-2919
USA

Cengage Learning is a leading provider of customized learning solutions with office locations around the globe, including Singapore, the United Kingdom, Australia, Mexico, Brazil, and Japan. Locate your local office at: **international.cengage.com/region**

Cengage Learning products are represented in Canada by Nelson Education, Ltd.

To learn more about Delmar, visit **www.cengage.com/delmar**

Purchase any of our products at your local college store or at our preferred online store **www.cengagebrain.com**

Printed in Canada
1 2 3 4 5 6 7 16 15 14 13 12

DEDICATION

For Viola and Ercole

TABLE OF CONTENTS

PREFACE

Introduction: A Textbook for Plant People

The spaces where people live, work, and relax must have plants present to enhance well-being. Indoor environments not having plants seem lifeless, hard, and artificial. In a fundamental way, plants bring life to the indoors. They soften plain walls, provide a sense of finish, reflect refinement, and improve work quality.

More than ever, consumers are aware of the beauty plants impart and how their use can improve interior surroundings. Potential interiorscape clients are constantly reminded of the beauty of plants and flowers in the media, seeing how they are used and displayed. Plants, from single specimens to complex installations, are highlighted in magazine photography, television shows, and Internet sites. Unique installations can be experienced and enjoyed in upscale hotels, shopping malls, office buildings, and other attractions. From large trees to delicate tabletop displays, plants are in the fashion forefront of interior design. Simply put, there are too many reasons not to have plant displays indoors.

This text was developed for students of both the art and science of plants. Today's new horticulturist must embrace the ways plants improve physical surroundings because this is the chief reason why plants are sold in the first place. Healthy foliage plants improve corporate image, making the clients of corporations feel well served. Professionally presented plants in elegant containers and contemporary design are as important to an interior as a painting or sculpture, so students should see them as such, gaining both curatorial and entrepreneurial skills along the way. This consumer-based recognition builds appreciation and enhances learning about plant growth and development.

Courses in home horticulture and interior plant identification and production are finding a renewed interest on campuses as students see the possibilities of growing and displaying plants in diverse, creative ways. Plants can be displayed on the floor, suspended from the ceiling, and adorn walls in varying proportions, from a simple wall planter to an entire wall of thriving, tropical plants.

This text is appropriate for courses centering on plant appreciation and growing and enjoying plants at home; courses where tropical plant growth and production are stressed; and courses where instructors appreciate the importance of horticulture as a business.

Professional horticulturists and interior plantscapers can benefit from this book to learn the basics of indoor plant display and care. Interior plants find their way to the consumer through floral departments in food and home building supply stores, garden centers and shops, and retail florists, so employees of these businesses will find this text indispensible. Rich with hundreds of photographs, *Interior Plantscaping: Principles and Practices* will spark the imagination of its readers and bring to life the practice of keeping plants looking their best.

How This Text Was Developed

I wrote this text to fill what I see is a gap in horticulture student learning: that a clear explanation of the principles and elements of interior plantscaping design is missing from their curricula. The media is full of examples of good design, but design principles are often forsaken for longer explanations about topics covered in other courses. Consequently, students learn only about the science of horticulture, not the artistic, business, and science topics.

As a horticulture student, I found that even though I was in a different horticulture course, text after text seemed to cover the same information. Soil, light, water, and other subjects were repeated, but the excitement of design and the challenge of business were, at best, a brief mention.

My first job in horticulture was at a plant shop and nursery during the plant boom of the 1970s. We are seeing another plant boom today, but it is different from that of the 1970s, with more sophistication and less repetition. The marketplace is ripe for new products, new ways of thinking that allow plants to be artfully displayed in the places where people work, live, and relax.

Today's student needs to see interior plants not only from the science of horticulture but as the art of horticulture. Plants fill a need for providing life to the indoor environment, but they must also be viewed as integral and necessary to interior design. There should be a planned, appropriate relationship between the plants maintained in the space and its interior design elements. It is important for an interior plantscaper to be as familiar with plant growth and development as it is to be familiar with the principles of design; after all, interior horticulturists bridge the gap between the outdoors and the interior environment.

This text embraces the culture of indoor plants in a tasteful manner. It is difficult to find publications that explain this broader view of plant culture indoors. Few sources appropriate for collegiate learners focus on plants as decorative accents for interior spaces. In the past, the approach to interior plantscaping has been to address the science of exotic plant horticulture alone. Previously published textbooks provided depth to understanding the effect of light and water on plant physiology, but they were short on design principles and business, two subjects sorely needed by horticulture students to be competitive in their careers. This text addresses the science of plants, but also covers them as indoor design elements. It relates the elements of interior plantscaping to the principles of design. The components of a plantscaping company and types of ownership are clearly explained. These major divisions of the text make this publication unlike any other available before. All of these topics are held within the context of maintaining the delicate balance of keeping plants looking good for as long as possible.

Textbook Design Organization

Interior Plantscaping: Principles and Practices is divided into four major sections. The first section is a series of chapters orienting readers to indoor plants. This beginning section, "Plants in the Interior," opens the mind of the learner to the possibilities of plants, growing them as a hobby all the way to creating diverse displays as a plant care professional. Chapter 1 encourages students to see indoor plant culture as a community of people who love plants, some who have emotional ties to them while others take a different route, becoming professionals in plant installation and care.

In Chapter 2, students are invited to see plants as design elements, considering them through the eyes of today's consumer. This is accomplished by challenging them to view diverse sites as places in need of interior plants. Chapter 3 is a key to the philosophy of the text, that plants must be viewed as living, decorative objects within a space and maintained in healthful balance. Readers will learn about the native habitats of plants and how they are named and classified in Chapter 4. This section concludes with a chapter on interior plant history, not only enlightening readers on the subject but also providing inspiration for modern forms of plant display.

The second section of the book is devoted to design principles and elements. Readers of the text are provided with foundational explanations of the principles of interior plantscaping design. Numerous, clearly presented examples help readers to understand and apply principles enabling them to communicate with designers and architects. Readers are reminded that much practice is needed to achieve levels of success in design and to remain humble. Chapter 7 introduces the elements of interior plantscapes similarly to the way floral designers or landscape architects view them, as ingredients of design. This section concludes with a chapter on complementary products and services including permanent botanicals, holiday décor, and floral design.

The science of horticulture segment begins with a discussion on the life cycle of ferns, the differences between gymnosperms and angiosperms and between perennials, biennials, and annuals, relating these plants to specific types used in indoor displays. Chapter 9 contains information on how plants can improve indoor air quality, including explanations about sick buildings and volatile organic compounds. Chapter 10 provides clear explanations of photosynthesis and respiration, and it includes a complete discussion of light energy. In Chapter 11 these topics are then related to practical interiorscaping, how to maximize the amount of time plants remain beautiful. In this section, further data covers the concepts of water, temperature, media, and nutrition. In Chapter 12, major insect pests are identified and earth-friendly controls are stressed including avoidance and the removal of pests, plant parts, or the entire plant when necessary. This section finishes with a chapter on plant diseases, both biotic and abiotic, and how to avoid problems by understanding the disease triangle.

The final chapters of the book focus on business aspects of interiorscaping, describing the various positions and duties within a company and how to start a plantscaping company. The techniques of plant care and display are described in Chapter 15, detailing watering, potting, cleaning, and other numerous topics.

Finally, the text concludes with a thorough glossary and an extensive appendix of popular indoor plants, their origins, care, and classical Latin pronunciation, many with photographs.

Features

Features in the book designed to make it more useful to the learner include:

- Green Tips written by students for students, this section offers chapter-appropriate suggestions for being more environmentally responsible in your plantscaping practice.
- Numerous full color photos aid in explaining abstract concepts, with particular focus on photos to illustrate design principles.
- Glossary terms are highlighted in each chapter, with definitions provided in the end-of-book glossary, to help make new or challenging terms more accessible.
- Plant appendix of over 125 plants for interior display. Many of the plants are considered interior plantscape staples, while others are more tender and appeal to the houseplant connoisseur. This section provides a phonetic pronunciation of the plant's scientific name in classical Latin, its common and family names, and a description of the plant, followed by maintenance, propagation, pest and disease susceptibility, and notes designed to help the learner distinguish and remember the plant.

Supplements

Interior Plantscaping Principles and Practices Instructor Resource CD-ROM

ISBN 13: 978-1-4354-3965-8

The Instructor Resources CD-ROM has three components to assist the instructor and enhance classroom activities and discussion.

Instructor's Guide

An Electronic Instructor's Guide provides excellent tools to help the instructor create a dynamic and engaging learning experience for the student. The Guide contains the tools listed here but can be downloaded and modified to meet individual instructional goals:

- Discussion Topics: These excellent strategies for stimulating class discussion can be used to challenge student critical thinking and create an interactive classroom experience.
- Additional Assignments: These ready-to-use activities and assignments are tailored to provide thought activities directly supporting individual chapters.
- Teaching Tips: This section provides engaging ideas and strategies for the instructor to use in conjunction with key chapter topics.
- Quizzes: These quizzes test students on retention and application of material in the text.

Computerized Testbank in ExamView™

- Includes 450 multiple choice, matching, and short answer questions that test students on retention and application of material in the text.
- All questions provide the correct answer.
- Instructors can create custom tests by mixing questions from the 15 chapters of questions, modifying the existing questions, and even adding additional questions to meet individual instructional goals.

Instructor Slides Created in PowerPoint

A comprehensive offering of over 280 instructor slides created in Microsoft PowerPoint® outlines the concepts from the text to assist the instructor with lectures and presentations.

ABOUT THE AUTHOR

Jim DelPrince is a professor at Mississippi State University (MSU). MSU's Floral Management program blends the art and science of horticulture together as a major, and it is home of The University Florist, a learning/teaching business operated by horticulture majors.

He holds an Associate of Applied Science degree in Floral Design and Marketing Technology, The Ohio State University Agricultural Technical Institute; a Bachelor of Science degree in Horticulture and Masters of Science in Agricultural Education from The Ohio State University; and a PhD in Agricultural and Extension Education, Mississippi State University. In addition to these degrees, his first job was working at a plant shop/nursery while in junior high school. In high school, he began a retail floral business in the basement of his parents' home. Since that time, he has worked for retail florists including Lew Kull Florists, Canton, Ohio, and Lane Flowers, Columbus, Ohio. He is a member of the American Institute of Floral Designers (AIFD) and Professional Floral Commentators International (PFCI).

ACKNOWLEDGEMENTS

I would like to extend special thanks to my mentor and friend Ralph Null, MS, AIFD (Fellow), Professor Emeritus, Mississippi State University, and to my colleague Lynette McDougald, MS, AIFD, Mississippi State University. I would also like to thank the students who wrote Green Tips and assisted with photography in this text: Tim Finnegan, Sarah Gordon, Kate Huseman, Chris Jones, Jordie Keffer, Jordan Kuhn, Aden Lunceford, Beth McDougald, Ellen Shivers, James Roy Sims, Morgan Skrmetta, Jessica Thomas, Tiffany Turner, and Ashley Vaughan.

The University Florist, Mississippi State University, the little bit of heaven on earth, supplied plants, containers, and constant inspiration for this work.

During my tenure at Mississippi State, I have enjoyed working for three permanent department heads—Richard Mullenax, Michael Collins, and Mike Phillips. Thank you for recognizing the importance of floriculture.

To the editors of this project, Juliet Steiner, David Rosenbaum, and Christina Gifford, thank you for your patience and sharing your expertise in writing and the world of textbook publications.

Special friends in the floriculture industry aided the development of this publication in many ways. Joe and Carol Gordy, NDI, shared beautiful photography of their high-quality floral designs. J. Schwanke offered much encouragement along the road; Steve Deason, Deason Wholesale, Inc., Kosciusko, Mississippi, provided live plant specimens; Barbara Helfman and Benay Leslie visited me and conversed about the industry; Sandra Bowdler hired me in my first job, selling indoor plants, along with Connie Palm, at The Happy Cricket in Ashtabula, Ohio.

I have been most fortunate to learn from some terrific university professors. Gary Anderson, PhD, and Mary Ann Frantz, MS, The Agricultural Technical Institute, Wooster, Ohio; Harry Tayama, PhD,

Tim Prince, PhD, and Rosemarie Rossetti, PhD, The Ohio State University; and Bill Parrish, PhD, and Jacque Deeds, PhD, Mississippi State University.

Feedback

Email address: jdelprince@pss.msstate.edu
James M. DelPrince, PhD, AIFD, PFCI
Professor
Mississippi State University

SECTION ONE

PLANTS IN THE INTERIOR

© Photographer's name/www.Shutterstock.com

The placement of plants indoors opens a world of opportunity for creativity, environmental enhancement, study, and appreciation. Plant displays are a must because they are living things and make people feel more comfortable. In order to work with plants indoors, horticulturists should be familiar with their names, their ancestral habitats, and how they were used throughout history.

CHAPTER 1

Keeping Plants Indoors

INTRODUCTION

Houseplants provide a wonderful way of bringing the outdoors to the interior. Their intricate patterns, interesting leaf forms, and pleasing proportions make them coveted accessories for any room in a home or apartment. Their presence on a table or windowsill adds a sense of life to the room and creates a cozy environment (Figure 1-1). There is a sense of excitement that interior plants bring to the viewer; nature has been tamed and brought indoors, closer for our inspection and pleasure (Figure 1-2).

FIGURE 1-1: A pair of *Aglaonema* plants, left in their grow pots, are displayed within a glazed, ceramic **footbath**. Spanish moss conceals the plastic grow pots.

Indeed, it is necessary for humans to feel closeness to nature, to draw it toward them. Some scientists refer to this inborn love of nature as **biophilia**. For example, it is not uncommon for people to keep pets at home. Many want to bring them to work, too, which is not always the best thing to do given the erratic behavior of the animal kingdom. It is not fair to suddenly introduce a pet to an unfamiliar and potentially frightening place; it also jeopardizes the health and safety of co-workers. What does work consistently is to bring the beauty of the plant kingdom into the home or any other environment occupied by humans. Plants soften the interior and provide a sense of calm and stability. Healthy plants presented in quality containers provide a positive image in a corporate environment (see Figure 1-3). Isn't it interesting that they are adaptable to a new environment so that they can be brought indoors, but were never meant to carry themselves indoors in the first place? They do not have feet, yet these immobile entities are moved and somehow survive!

People manipulate plant materials all the time. We dig field-grown trees and use them for landscaping, cut flowers from greenhouse-grown plants for floral designs, and harvest edible parts for salads. What many people do not think about is the way the plant kingdom manipulates us.

FIGURE 1-2: A fountain of plants creates a focal point in the Embassy Suites Hotel, Baton Rouge, LA, hotel atrium, near a restaurant entrance.

FIGURE 1-3: The contrast of lush greenery visually frames an elegant staircase while a single plant provides a buffer for an interior column.

Plants do not have brains to think, yet they have evolved and changed over the millennia to become attractive to humans, thus perpetuating their species. Plants provide us with food through edible roots, stems, **petioles**, leaves, and flowers. People have evolved with plants, seeing beauty in their leaf forms and fragrant, colorful flowers. We take care of plants because, in their own way, they take care of us. Ornamental plant materials have the principles of design built-in, providing a set of reasons of why they are so beautiful (see Figure 1-4).

FIGURE 1-4: Plants provide a subject for both art and science.

© wavebreakmedia ltd/www.Shutterstock.com

FIGURE 1-5: By caring for plants, people can learn nurturing skills and feel a sense of reward.

In this way, plants become pets of sorts for humans, and the **people/plant interaction** becomes mutually beneficial. A person cares for a plant, provides it with water, sunlight, and appropriate temperature. The plant provides beauty as long as it remains healthy. There are many more things that ornamental interior plants offer, and they will be covered further in the text, but there you have it. Plants are beautiful and we are their pawns (Figure 1-5).

Science and Design, from Hobby to Career

Growing plants at home is a terrific hobby due to the huge range of varieties available. Some people like to specialize, growing just one type of plant and exploring its world of variation. These people are very involved in their personal people/plant interaction and have earned the enviable right to be called "plant geek." Their affinity for plants has a noble quality due to their devotion to a family or genus. Others like to maintain a variety of plants and use them as design accents, for instance, a commanding specimen palm in a contemporary setting or a delicate fern in a mossy pot as an accent on a bedroom table. Much of this style of plant usage is owed to a refreshed way of looking at interior design, brought about by magazines, television, and Internet information on interior design. Artifice is to be avoided while things that are natural and alive are revered. If the design accent dies, it is replaced with another plant or a book or lamp. What is important to them is the "look" of the space. This underscores the importance of interior plant *design* (Figure 1-6). We need to work well beyond the simple delivery of a

FIGURE 1-6: A pair of planters was specified by an interior plantscape designer as necessary in this space to bring pattern and color to an entry.

post-production, potted plant to an interior space. We must scrutinize the plant within the context of the principles of design.

Yet another type of use is in the culture of a person with emotional ties to a particular plant. It is not uncommon to see plants that are kept, whether in robust condition or barely hanging on, because they were a gift for a birth, a birthday, or a funeral. The owner simply cannot part with the plant because it is a constant reminder of a lifetime event or a special loved one. Throughout all of these, the constant theme of nurturing comes to play. If the plant is cared for, it survives, at least for a time, rewarding its owner with its beauty (Figure 1-7).

FIGURE 1-7: An ivy topiary plant is accented with a plush toy, thus creating a themed gift.

Retailing of plants and services is a profession within interior horticulture. The provision of plants for income

FIGURE 1-8: A balance of knowledge, skill, and business acumen enables this indoor display to exist.

expands the responsibilities of the horticulturist. The necessity of providing and maintaining healthy, disease- and insect-free plants, offering competitive pricing and maintenance regimes, operating a business that consistently generates profit, and working with employees and clients introduces important challenges that require a variety of skills. Professional interiorscaping involves the profession of planning, design, installation, and maintenance of living plants (Figure 1-8).

The people who manage interiorscape firms must meet many challenges beyond selecting and maintaining the right plants. They are required to be skillful in the world of business, possessing knowledge of accounting practices and marketing. In the realm of horticulture, they should know and appreciate design principles as well as plant science. Balancing employee relations and clients, both existing and potential, they teach others about the benefits and care of plants while respecting the thoughts and feelings of others (Figure 1-9).

In cities where business can be supported, an interiorscape department within an existing horticultural enterprise should be considered. For example, a logical outgrowth of a garden center or retail nursery would be to sell and maintain tropical plants for interiors. This could be started slowly, such as in the rental of plants for the short term or special events, where the garden center delivers display-quality plants to a venue, places them appropriately, then picks them up after the event is completed. Over time, rentals

FIGURE 1-9: Interiorscapers enjoy working with plants and the community. It is a rewarding business, bringing the beauty of nature to people.

become long-term, which could then involve revenue-generating maintenance contracts.

Plants and associated materials are attractive to people because they ultimately improve the look of the home and workplace, thus supporting a better lifestyle. A product mix of interior plants, exterior/landscape plants, and interior design accents complement each other in areas where there is a demand (Figure 1-10). In this way, complementary products highlight and sell each other.

Many successful start-up companies were developed in the homes or garages of their first owners. In general, office space for bookkeeping and communications and compact storage space may be all it takes some companies to start and grow their clientele list. There is no need for board rooms or consultation space when this service is conducted at the client's location. Since most interiorscapers do not produce their plants, greenhouse space may not be needed if the product can be turned around quickly.

FIGURE 1-10: Attractive home accents and gifts widen the product offerings of horticultural retailers.

The results of keeping plants in a home or professional space not only enhance the physical environment but also go further to uplift the feeling people get in the space. Healthy plants add a sense of flair and provide evidence that the home or business owner is thoughtful and nurturing about the way things look and the way that guests feel welcomed. Employees and clients of companies with well-maintained interiorscapes feel more comfortable while in these professional environments. When horticulture professionals are hired not only to install but also to maintain plants, it becomes obvious

© Neon Light/www.Shutterstock.com

FIGURE 1-11: A natural, green interior makes people want to linger and enjoy their surroundings.

that the company is thoughtful about the way it appears to its employees (Figure 1-11).

Keeping plants indoors blends design and science unlike anything else. The benefits that plants offer surpass other design and mechanical elements that can be introduced to a living or working environment.

The world of interior plantscaping is more than just a hobby. People's need for keeping living things close by provides the reason why interior plantscaping exists as a profession. Within the organization of an interior plantscape company, many talents are required, especially that of the plant care specialist.

The goal of this text is to increase knowledge and appreciation of the challenges of professionally maintaining plants indoors. There is much to learn about the science and art of indoor plant culture, for interiorscaping directly involves both.

Interiorscaping answers the need for creative people to create. Many people search for careers that reward them with a sense of satisfaction, jobs where they can work with people and beautiful things. Some people love detailed, intricate work while others may enjoy the installation of large-scaled landscapes using forklifts and numerous crewmembers. Interior plantscaping answers to both types of people. It is landscaping performed and displayed within the great indoors.

Interior plantscaping is curatorial. Some interiorscapers are true connoisseurs of botanical specimens and prize the rare and exotic. An affinity for learning the origins and breeding of plants can lead to careers with botanical gardens or to public or private collections.

GREEN TIP

Throughout this text, you will find boxes of information called Green Tips. They are meant to make us think about reusing, recycling, or repurposing materials within the realm of interior plantscaping.

Necrotic foliage can be removed from indoor plants and added to a compost bin located at the interiorscape home location. Though not recommended for interior plant use, well-rotted compost is one of the best amendments for outdoor gardens.

Image Courtesy of Gaylord Opryland Resort & Convention Center

FIGURE 1-12: Professionally installed and maintained plants decorate and articulate space.

The profession of interiorscaping aside, it is still fun to work with plants. With a small amount of careful maintenance, it is possible, and often quite simple, to sustain a captivating piece of nature. This reward holds true from a single plant to acres of plantscaped interior displays (see Figure 1-12).

SUMMARY

People have an intrinsic need for nature, and plants are a convenient way of surrounding ourselves with nature. These products of nature are living decorative accessories, as important as furniture, lamps, and art, and sometimes take on emotional significance as well. It takes many skills in order to operate an interiorscaping business that can be initiated within an existing, related establishment such as a garden center or nursery or that can grow as a small business with limited initial capital.

CHAPTER 2

Plant Uses, Plant Opportunities

INTRODUCTION

Chapter 2 provides perspectives on numerous opportunities for the installation and maintenance of professional interior plantscapes. Some sites might be considered common venues for plantscapes such as office buildings and hotels (Figure 2-1), while others are more unique like zoos (Figure 2-2) and schools. Note that all interior environments would be enhanced by the addition of plants, not just those listed. The challenge as a plantscape professional is to constantly pursue possibilities for plant placement sites. At the crux of this pursuit is to find clients with available funding and to educate and persuade decision makers that interior plants are important for ambience and human well-being.

FIGURE 2-1: Multi-floor hotels with open atriums benefit from the inclusion of plants.

Seeking out potential opportunities for professional interiorscapes is demanding but fun and is accomplished through contact with people, observation of spaces, and exploration of potential sites. Good interiorscaping involves embracing good sales techniques; the basic tenet of marketing "find a need, fill a need" remains true. Owners of interiorscape companies are motivated to find new clients in order to generate necessary revenue to operate a business and pay associated expenses. All too often, employees are not driven, their needs motivated only by acceptance of a regular check. In today's world of work, employees working in interiorscaping should

FIGURE 2-2: A Tiger Python feels and appears more at home in its native tropical habitat when surrounded by plants such as *Epipremnum aureum*, Devil's Ivy.

consider how they can make themselves valuable to their company. One way of building employee value within the company is to find and secure accounts so that their employers would continue to hold them in high esteem.

Horticulture professionals cannot hope to be successful in business by waiting for the phone to ring. It is necessary, and fun, to get out of the greenhouse and meet people. After all, it isn't always what you know but who you know. People are most always interested in learning more about plants, their care, native habitats, flowering, durability, and more. Sometimes, when people possess knowledge of a certain area of study, they have the attitude "Everyone knows that!" Perhaps they feel as though others are not interested at all. This is simply not true. Students in horticulture have multiple interests and talents; capitalizing on strengths can add up to successful careers or hobbies that nurture and create a happier lifetime experience.

Plants are used for aesthetic purposes, it's obvious, but the thousands of plants available on the market suited for interior culture multiplied by the millions of combinations, containers, and other forms of presentation mean that visitors will never tire of them as long as they are thoughtfully designed for a space. Good designers rarely if ever create an exact copy of a design. When exact designs or specific plants are repeatedly used, they become mundane. When people tire of something, they often abandon it. This is not beneficial for individual businesses nor is it good for the industry as a whole. Variety is important, and designers can find inspiration from plant materials and containers (Figure 2-3).

FIGURE 2-3: This Cycad at the Atlanta Botanical Garden possesses a blue-green tint, a similar value evidenced in the **jardinière** holding an *Aloe*. Because this setting has bright, natural light, unusual plant choices were specified.

© Capital Atrium, Belle of Baton Rouge Casino & Hotel, Baton Rouge, Louisiana

FIGURE 2-4: A *Rhapis excelsa* of the appropriate height and width was specified and installed in this semi-circular planter. It softens the support surround, yet does not interfere with the light fixture.

Plants can be used in many places; this allows creative horticulturists to specialize in numerous areas of tropical plant culture. The notion of **site specificity**, meaning that a certain plant or plants and their containers are specially chosen for an exact site, makes each design situation a little different (Figure 2-4).

There are many, many types of public spaces that greatly benefit from well-designed, well-maintained interiorscapes. The segments in the following sections provide possibilities for interiorscape business prospects as well as the possibility for part-time or full-time employment.

Adult Living Centers

The beauty and restful nature of indoor plants provides a soothing and natural environment in which to live (Figure 2-5). Senior residences benefit from communal areas accented with plant displays. Such an environment provides therapeutic activity to watch plants grow or appreciate the addition of seasonal color. Although most of the plants that withstand the rigors of interior conditions such as low light and limited watering do not frequently, if ever, offer fragrant flowers, introductions of plants with scented foliage and flowers can spur delightful memories in the elderly. It is worthwhile to include such bonuses in the monthly plant budget,

© GWImages/www.Shutterstock.com

FIGURE 2-5: Settings such as this bring nature indoors to those who are less mobile.

© Halina Yakushevich/www.Shutterstock.com

FIGURE 2-6: Residents within assisted-living facilities would love to take care of plants if given the opportunity.

especially during the gray days of winter. Regions with extreme cold in winter or with hot summers should have interiorscapes to allow for year-round gardening in comfortable temperatures.

Often, residents are invited to participate in the regular maintenance of interior plants when interiorscaping is accomplished in-house, meaning the installation is overseen by trained employees of the home. Such people/plant interaction is of great advantage, allowing residents the opportunity to garden close by, in the safety of an indoor environment. Tasks involved with the care and maintenance of plants encourage standing, stretching, bending, and reaching, which are beneficial for those in need of gaining and maintaining flexibility. They also encourage socialization of people with similar interests in gardening and the natural world (Figure 2-6).

It makes good sense to involve center residents in horticultural tasks. Many of them may miss their home gardens, so taking care of plants can stimulate memory cognition. Staff horticulturists can assist with maintenance preparation, heavy lifting, and other tasks associated with horticultural supply management.

Often, a hindrance for starting and maintaining horticultural therapy programs is funding. Administrative staff would like to improve resident services, but in order to keep residents' costs in line, activity expenses are often the first to be cut. For those seeking employment or service offerings in such facilities, the benefits of interior horticulture should be highlighted. Discussions should center on not only the physical and emotional benefits to residents but also the way a professionally maintained interiorscape improves the space and makes a better impression upon visitors. Some facilities do have the financial means necessary to employ staff horticulturists and horticultural therapists.

FIGURE 2-7: Botanical gardens, such as the Atlanta Botanical Garden, maintain a living record of exotic plant materials.

Botanical Gardens

Students pursuing degrees in ornamental horticulture are in demand for jobs, internships, or volunteer positions in botanical gardens where tropical plants are maintained (Figure 2-7). Botanical gardens bring important missions to the larger community through the maintenance/preservation of collections, learning activities, and research. Visitor expectations run high when observing these collections. Careful maintenance of rare, seldom-seen plant varieties is a hallmark of horticulturists stationed at botanical gardens. Additional opportunities are available in display gardening where exhibits are rotated according to specific themes or seasons. Staff horticulturists pay close attention to focal areas such as conservatories, reception areas, and administrative offices.

Casinos/Gaming Facilities

High-profile casino sites are often laden with lush plantings and distinctive seasonal holiday displays of live plants and related décor. The more alluring the space, the more guests will be attracted to visit and patronize a casino and its various amenities including hotel, restaurants, spas, salons, and shopping. Interior gardens provide theme and ambiance to the atmosphere, thus widening appeal to varying clientele. In better establishments, horticulture management plan displays one to two years in advance in order to have crops specially grown for the casino. Hotel lobby space, gaming spaces, and themed restaurants all benefit greatly by the use of varying plant materials to support themes and carry a sense of welcome to patrons. Every visitor can enjoy the sense of being a wealthy, "high roller," if just for a few hours.

Educational Facilities and Institutions

Unfortunately, environments that should foster creativity are often the most neglected. Spaces frequented by students, faculty, and visitors at schools, colleges, and universities need lush plants placed in focal areas (Figure 2-8). Research with plants in professional work environments points out that higher productivity levels exist when plants are introduced and maintained. Beyond efficiency in business, this finding can be applied to enriching the work of students in K–12 and higher education. It is important to introduce young people to the benefits of plants so that they continue to seek them out and enjoy them throughout their lives.

Healthcare Facilities

Environments where people feel stress can be improved with the addition of professional plantings. Foliage plants introduce a feeling of tranquility and serenity. Patients waiting on appointments at an ophthalmologist's or psychologist's office or visitors in emergency room waiting areas should be exposed to the benefits of verdant, green plants (Figure 2-9). The many shades of green plants are peaceful; thus, the built-in psychological effect of a mass of foliage plants aids in helping people to feel calm.

Plants may help people to endure pain for longer periods of time. Similar to a hospital room's artwork or window view, plants can take the focus away from pain and make the space feel fresher.

In the past, some hospitals would not allow certain plants or flowers in patients' rooms for fear of introducing pathogens harmful to human health. Some would cite potential problems such as allergies caused by pollen. Few of the durable plants used for interior plantscaping bloom prolifically indoors. Those that do offer blooms in low light intensities (*Spathiphyllum*) or continue to hold floral spikes for long periods of time (*Aechmea*, Orchids) bear pollen too heavy to be airborne indoors. Today, we know that plant diseases are neither transferable nor harmful to people and that the addition of plants to a room makes people feel better. The psychological uplift people receive from a gift of plants or flowers

FIGURE 2-8: Green plants give students a nature break after spending numerous hours working indoors.

FIGURE 2-9: This waiting room would appear ordinary and austere without these plants.

outweighs the cost. Indeed, a growing body of research points out that plants improve indoor air quality, an issue that will be further explored in Chapter 9.

Hotels

Plant adornments are a natural to decorate a hotel. Hotels offer hospitality to their guests in many ways, and a large part of the welcome is providing a pleasing environment with attractive interior design. Plant displays in such places serve many purposes. They can be accents used to introduce a natural, earth-friendly component in lobbies. Architectural elements like load-bearing columns and **pilasters** may be visually softened with fine-patterned foliage plants. Visual screening can be accomplished with a filmy panel of foliage plants in a single planter or in individual planters grouped together. Visual, rather than physical, separation of administrative or service areas does not obstruct vision, but provides a living curtain of greenery (Figure 2-10).

Malls/Retail

Some of the most impressive interiorscape designs ever created have been installed in shopping malls and other retail spaces. Colorful plants and flowers attract attention, and smart vendors have used plants as a marketing technique long before the shopping mall concept was conceived. Retail centers become more of a destination when they feature noticeable plant displays. Commanding plantscapes cause shoppers to slow their pace, thus creating greater opportunities

FIGURE 2-10: Plants' natural patterns harmonize with architectural lines.

for them to see items in stores, turning interest into a desire to buy. Plants improve real estate value, enabling mall owners to gain higher rental charges per square footage. In cities with multiple, competing shopping destinations, periodic upgrades to existing interiorscapes provide a noticeable lift to shopper and tenant satisfaction. The appearance of well-kept foliage and flowering plants signals an upscale shopping experience, creating excitement before the shopping experience even begins (Figure 2-11).

© Beau Rivage, Biloxi

FIGURE 2-11: Seasonal color displays as seen with these Hiemalis *Begonias* signal an exclusive shopping experience.

Interiorscape design for shopping malls has evolved over the past 40 years, with more importance placed on style and selection rather than merely placing ordinary plants in standardized planters. Plant material choices have moved from using just a few varieties of dark green foliage to using new foliage plant varieties, bromeliads, and orchids (Figure 2-12).

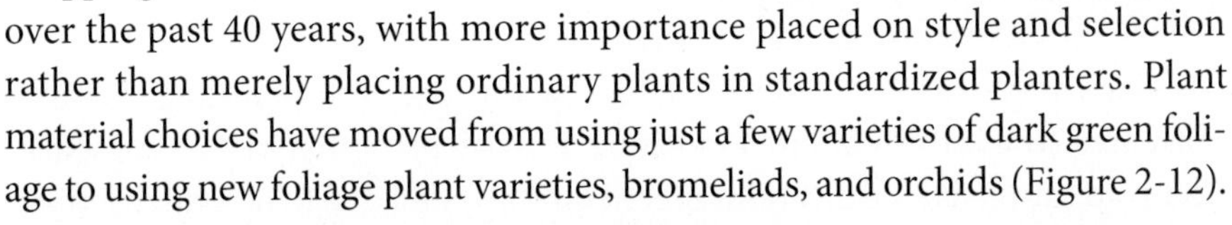

© Cengage Learning 2013

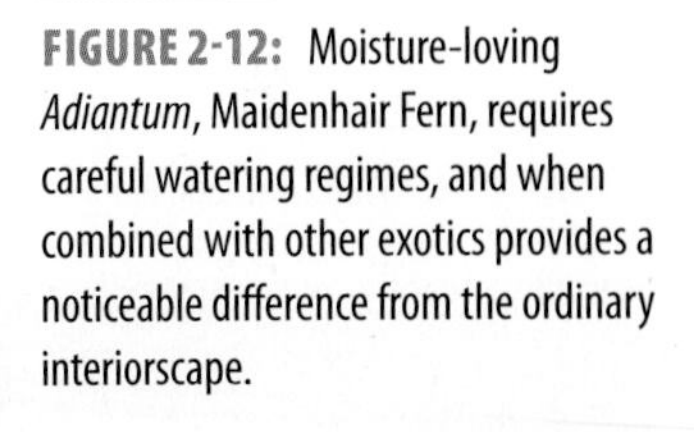

FIGURE 2-12: Moisture-loving *Adiantum*, Maidenhair Fern, requires careful watering regimes, and when combined with other exotics provides a noticeable difference from the ordinary interiorscape.

FIGURE 2-13: Masses of Poinsettias, *Euphorbia pulcherrima*, on display at the Atlanta Botanical Garden are a holiday classic.

Another important part of the evolution is the increase in demand for holiday decorations. Who better than a talented interiorscaper to design, install, and carefully remove seasonal decorations throughout the year?

In keeping with holiday retailing promotions, interiorscapes can adapt to the season reflecting marketing approaches for Christmas, back to school, Easter, and Halloween (Figure 2-13). Colors evocative of the time of the year as well as seasonal plants may be incorporated into existing interiorscapes. Plant care companies are best at providing additional props and scenery for Christmas and Easter exhibits, housing Santa Claus or the Easter Bunny. In this way, entire small-scale buildings and landscapes add to holiday excitement and draw visitors for photography and shopping. These seasonal displays require intensive design, installation, dismantling, and storage, but they can be a significant revenue generator for interiorscape firms.

Museums

Lobbies and restaurant spaces in museums benefit from the installation and maintenance of sculptural green plants and floral designs. Tasteful dish gardens and blooming plants provide a sense of welcome at information stations and restaurants. Large scale plants, for example, regal Kentia Palms, should be used to add nature's grace to entries.

Care must be taken in the marketing approach for plant provision in museums. The introduction of moisture and live plant material in the spaces occupied by museum collections is generally avoided due to potential damage to artworks. Although plants may provide a desired aesthetic, even small amounts of moisture may introduce a missing key that will aid in the development of mold, mildew, or other harmful pathogens that go beyond damage to plants. It is one thing to lose plant health, but imagine the harm that could be done to priceless paintings, tapestries, or other media? Most of the time, damage is caused by careless watering. An unwitting plant technician may slosh water causing it to hit walls and floors. If a plant container is not waterproof and there is nothing to catch water that has percolated through the pot, moisture damage to surfaces will occur. This can cost the interiorscaper hundreds or thousands of dollars in repairs and replacements. It will also break down any established goodwill between the client and the interiorscape firm (Figure 2-14).

FIGURE 2-14: Water damage to floor.

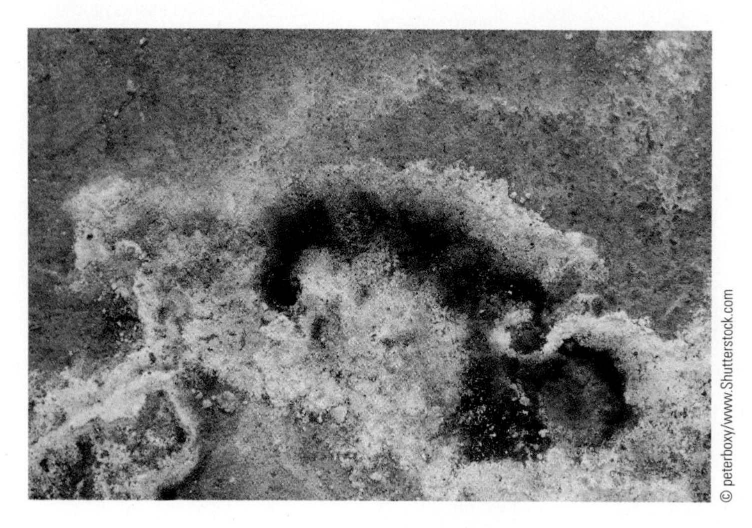

Curators often keep plants away from art collections due to the possibility of introducing *Dermestids*, common name Carpet Beetles. The adults of this insect are able to fly from room to room and the females lay eggs which develop into larvae. The larval form of the insect is the most destructive stage, consuming, among many other things, wool and silk, the media for tapestries and fabrics in period displays or sculpture. Indoor plants do not attract such insects, again because the larval stages do not feed on plant material, but this may be impossible to convince careful staff members overseeing priceless collections. It is better to be safe than sorry, keeping beautiful plant displays to the common areas where people congregate and keeping all stakeholders happy.

GREEN TIP

Artistic presentations of live plants provide "living art"; thus, an interiorscaper can provide one-of-a-kind wall hangings and three-dimensional sculpture using plant materials. Living art can be used in a space instead of paintings or prints.

Offices

Whether a large company or small, in times of economic boom or financial meltdown, corporate image is of great importance (Figure 2-15). Bank customers seek out institutions that protect their investments and conduct transactions in an expert manner. Clients of legal entities place their trust in lawyers while tax experts are expected to have full knowledge of all current tax laws.

Robust interior plants in well-designed containers are the perfect accent in corporate environments and institutions when provided by a

FIGURE 2-15: The entrance lobby of a multi-tenant office building should appear well-maintained and reflect the professionalism of its leaseholders.

© bendzhik/www.Shutterstock.com

FIGURE 2-16: A dying or dead plant sends negative messages.

skilled interiorscape firm. Updated designs and plant selections provide an appearance that the company is in-the-know when it comes to managing client needs. Conversely, weak, spindly plants give the appearance that the institution is in financial difficulty and trying to cut corners. In all situations, a few well-maintained plants are much better looking than several weak plants clinging to life, maintained by well-meaning employees (Figure 2-16).

Places of Worship

Long-term displays of lush green plants and sometimes flowering plants are important to the look and feel of large-scale places of worship (Figure 2-17). Congregations may build budgets for plant decoration; this is especially true with places of worship that hold activities throughout the week. In order to create a more attractive place that makes everyone feel welcome, it makes sense to have appropriate-scale plants in beautiful containers in lobbies, classrooms, and offices as well as sacred spaces. Consideration must be given to amounts of available light, including natural and artificial sources. Places of worship have members who continue to be a part of the space, often on a daily basis.

© Stacy Barnett/www.Shutterstock.com

FIGURE 2-17: A Christmas altar bearing Poinsettias at the Santa Barbara Mission, California.

Residences

Active lifestyles do not always allow the pursuit of home horticulture, although many potential clients have the desire for healthy, beautiful plant displays in the home (Figure 2-18). Busy professionals have the means to subscribe to plant installations and care services. More than ever, consumers are conscious of the importance of the home having the appearance of a showplace that could be ready for a photo shoot at a moment's notice. In order to maintain what seems to be an effortless approach to living, homeowners can hire a professional interiorscaper to provide them with important specimen plants. Such plants placed in focal areas of the home, especially but certainly not limited to living rooms, media rooms, kitchens, and baths, provide them with a sense of tranquility. Nearly every photograph in today's stylish shelter magazines features flowers and plants. These images underscore that plants and flowers steal the show from all other elements of interior design. An elegant, arching *Phalaenopsis* orchid plant in a striking container could easily be the first object seen when entering a room. Such a statement of good taste is desired by clients, but skillful, reliable service may seem impossible to source and maintain.

The effects of having one's own gardener can be found by employing an interiorscape company to install and take care of plants. Obtaining and keeping clients in this category takes social skill, horticultural knowledge,

FIGURE 2-18: An elegant living room featuring live and permanent plant displays.

© Erik E. Cardona/www.Shutterstock.com

FIGURE 2-19: This complementary color combination of red and green invigorates and refreshes dining patrons.

and a sense of design style often possessed by those who love plants and flowers.

Restaurants

Dining out has become a regular part of many people's experience, and destination choices abound. What sets many successful restaurants and bars apart from the rest is the atmosphere or ambience provided by floricultural displays (Figure 2-19). Poor displays with declining plants or dusty artificial plants may hint at bad service or food. A professional plantscape display can add to a restaurant's culinary theme, whether Brazilian, Moroccan, or Vietnamese. Implementing an effective maintenance program is just as important as choosing the right plants to support a theme or a feeling desired by the restaurant's interior designer or owner. Plants must be clean, free of greasy, dusty residues, and must not harbor yellow leaves or trash at the soil line. Dining patrons feel comfortable in an environment with a sanitary appearance.

Plants in restaurants aid in space articulation. Patrons can feel a sense of privacy when plants are used to segment and separate groups of tables. A dining experience is improved when clients can converse in a more intimate setting and are able to hear each other rather than conversations at other tables. Just as interior furnishing such as drapes, upholstered furniture, and rugs stop sound from bouncing around a room, plants, too, reduce noise levels. They absorb, refract, and reflect sound, so a more abundant approach to interiorscaping using larger planters would prove more effective.

Retail Establishments

Designer clothing stores, department stores, and other upscale retailers have unique spaces that would be enhanced with unusual plants. A plantscape designer can create an atmosphere that highlights the products and services of the client; after all, a good installation helps the client's image. Small businesses do not have as sizeable budgets to work with as do larger corporations, but are sometimes quite cutting edge in style and design. Such accounts challenge the interiorscaper to come up with unique designs on a tight budget. Smaller accounts are sometimes easier to manage and may be a great way of testing the waters of professional interiorscaping before tackling large installations where there is much to gain or lose. For retailers with ample budgets,

FIGURE 2-20: Placement of these colorful *Aechmea* help to draw attention to merchandise in a furniture showroom.

displays aid in getting the attention of buyers through the alluring patterns and textures plants provide (Figure 2-20).

Zoos

Ambience is necessary for quality living, not only for humans but also for primates and other vertebrates. An important part of keeping animals healthy is to simulate their native habitats. The provision and maintenance of plant materials that are similar to those found in the geographic locales of various species need to be balanced with what will flourish in the artificial habitat. Many zoos retain people with knowledge and skill of caring for tropical plants. Interior plants provide the visual accent aiding in suspension of disbelief that visitors are not in a zoological park but rather in a native habitat.

Green Buildings

Research and design efforts have been brought together to make entire buildings, new or already constructed, inside and out, better for the environment. A **green building** is one in which the design, construction, and operation of the construction is done in an environmentally friendly way. Throughout a green building's life, from site selection, design, construction, operation, maintenance, renovation, and, finally, deconstruction, environmental stewardship plays a leading role.

The United States Green Building Council (USGBC) created LEED as a rating system for green building. LEED stands for Leadership in Energy and Environmental Design. Part of the LEED philosophy is the involvement of indoor environmental quality. Frequently, interior plants are included in the design of green buildings due to the documented benefits they provide, including biophilia. Plants naturally reduce stress and increase creativity.

SUMMARY

Plants provide positive effects on building sales, promoting relaxation, and softening architecture. In short, people's lives are enhanced by well-positioned, professionally installed and maintained tropical plants. A focus on plant provision enables plantscapers to envision many more environments where plants could be placed, far beyond those mentioned in this chapter.

Interiorscaping as a career relies on design, sales, installation, and care. In order to accomplish those goals, we must see the possibilities of interior plant placement in numerous indoor environments. Start by practicing in places where you trade for products and services. Does your doctor's office leave something to be desired in relation to healthy, well-displayed plants? The first step to building an interiorscape business is to start dreaming of the possibilities.

CHAPTER 3

Cultivation versus Aesthetics

INTRODUCTION

Many people who enjoy indoor plants refer to the process as "growing" plants indoors, rather than just "keeping" them indoors. Their reference to the production of stems and leaves is correct because they know the necessities of plant growth, especially bright light. The soil in which they keep indoor plants provides an adequate balance of moisture, aeration, and nutrition. Such healthful combinations will result in plant growth. What is not necessarily considered is the maintenance of a particular level of aesthetics—the sense of the beautiful. While beauty is in the eye of the beholder, much more is at stake in the world of commercial design.

Display Duration

In keeping plants indoors, the most important thing that drives the philosophy of this text is plant aesthetics (Figure 3-1). The goal of keeping plants in the interior is general but very important: to provide beauty. It is not important to grow plants indoors, and, due to environmental limitations, it is not always possible. Note the title of this section, *display duration*, which gives insight into the notion that plant groupings are living displays. The interior plantscaper is a horticultural display designer, a human interface between the outdoor world and the great indoors.

It may be interesting to approach this subject with a few examples of what not to do. The following list provides examples of plants that possess poor aesthetics, which should be avoided with interior plants.

Don't use it when a plant

- requires staking in order to remain upright.
- has lost too many lower leaves resulting in awkward proportions.
- is top heavy and visually unbalanced in the container.
- has grown too large for its space.
- has infestations resulting in sticky honeydew.
- has a "Charlie Brown Christmas tree" appearance.

FIGURE 3-1: A magnificent *Tillandsia* at the Atlanta Botanical Garden keeps its good looks without soil, living on sunlight and humidity.

FIGURE 3-2: A plant that has outgrown its container and surroundings detracts from the interior design and the client's image.

FIGURE 3-3: *Epipremnum* are vining-type plants. They can be trained to grow upward, but they need tending in the process. Stems of this Devil's Ivy should have been attached to the support with non-descript **mechanics** rather than the bright green tape. As a result of too little training, the plant has lost its aesthetics and should be changed.

Of course, all of these plant problems and others can be rectified, but interiorscapers are not plant doctors nor is the client's space a plant hospital. The goal is not necessarily the production of new roots, shoot, and leaves. Sometimes growth of a plant in the interior is detrimental. A growing plant can outgrow its decorative container or the entire space that it was originally designed to adorn (Figure 3-2).

At this point, the owner of the plant may have trouble making a decision as to what to do. Continual staking, often with non-decorative materials such as a yardstick, rough, two-by-two lumber and string, or worse yet, recycled pantyhose cut into strips, makes a mockery of what was once a regal plant (Figure 3-3).

Low light situations are the chief reason why plants do not flourish, or grow at all, indoors. Seen in another way, the lack of light becomes a growth regulator, slowing down root and shoot initiation. It can be seen as beneficial because it keeps a plant from outgrowing its intended site (Figure 3-4).

FIGURE 3-4: *Philodendron* are some of the best-loved, low-light plants.

The length of time a plant remains aesthetically pleasing is subjective. Some people have cherished plants on display that others would have discarded years ago. Decisions are made when a plant is somewhere in the balance between healthy/aesthetically pleasing and unhealthy/sickly appearing. It is helpful to know that a dead plant is biodegradable. It is able to become compost, thus enriching to soils, so if it is thrown away, it will biodegrade in a landfill. The same cannot be said for plants made from vinyl, which degrade but do not biodegrade.

A thorough knowledge of plant families helps the interior horticulturist because if you know the care of one of a plant family's members, you will have a good start on the care of anything in the family. You will know which plants are best suited to the conditions at hand, thus helping to ensure the installed plants stay aesthetically pleasing for a long time.

GREEN TIP

A dying plant has lost its aesthetic appeal, but a positive spin on what might otherwise be a negative situation is that a plant is biodegradable. Knowing that dead plant parts will disappear and feed beneficial microbes along the way, clients will feel better about plant replacements.

The Native Habitat: The Plant's True Home

If we know the plant's native habitat, we will know its indoor care (Figure 3-5). Most of the interior plants that are available on today's market are from ancestors that grew and reproduced in warm, tropical climates. They were understory plants, sheltered from the intense tropical sun by taller trees that can tolerate extreme light levels. This is the reason why such plants perform well in the low light of interiors. The plant life growing under protective trees flourished in the dappled sunlight, warm temperatures, and frequent rainfall

© amrita/www.Shutterstock.com

FIGURE 3-5: Conditions mimicking the tropical rainforest allow *Heliconia* plants to thrive, thus producing its blazing colored bracts.

© Chris Curtis/www.Shutterstock.com

FIGURE 3-6: A tough terrain yields incredible beauty as seen in this *Agave* with its needle-sharp leaf terminals.

in the tropical forest. The abundance of plant and animal life created soil rich in decomposed organic material with ample water-filled and air-filled pore spaces.

Tropical forests might not have extremes of cold and heat, but fluctuations do exist which have made their mark upon the native plants of those regions. Plants at higher elevations could experience temperature dips around 50°F. Plants such as *Sansevieria* and *Cymbidium* can take these cool temperatures not only in their tropical homes but inside homes and offices where thermostats remain low during the night or weekends. This is not to say that they do well in cold, drafty areas such as entryways and vestibules. Short bursts of cold air can be detrimental to the health and display life of many types of tropical plants.

A plant's appearance will often dictate its geographic origins, such as the case with succulent plants like *Agave* (Figure 3-6) and *Aloe*. These plants have adapted here the ability to store water in their thick leaves. *Nolina recurvata*, the Ponytail Plant, has grass-like foliage in a thick tuft born upon a stem, but the stem holds a wide base that acts as the water-storage portion of the plant. Lacy *Nephrolepis,* Boston Ferns, take on a gray-green pallor when moisture-stressed, demonstrating their origin from the high rainfall regions in tropical Africa.

FIGURE 3-7: This miniature Bamboo plant is benefitting from its placement in a humid, brightly illuminated bathroom.

High rainfall results in high humidity levels, one trade-off that is sometimes difficult to achieve indoors. Plants can suffer from less-than-adequate indoor humidity resulting in brown leaf margins and tips. If problems persist, steps must be taken in order to raise humidity around the plant or relocate the plant to an environment with adequate relative humidity (Figure 3-7).

Arid deserts spawn a completely different set of plant characteristics. These plants are able to take up limited rainfall quickly and store it in their fleshy leaves and stems. Many families have adapted ways of protecting their water stores with prickles and spines to ward off predators. Deserts may be unbearably hot to humans during hours of sunshine, but they may also be very cold at night. Cacti and succulents are tough plants for tough climates that may be somewhat similar to conditions in domestic or commercial environments. Indoor settings with very high light intensities and pockets of heat with associated low humidity create the perfect setting for desert plants (Figure 3-8).

FIGURE 3-8: A miniature garden of *Crassula* and other succulents will perform well in strong light, low humidity, and little water.

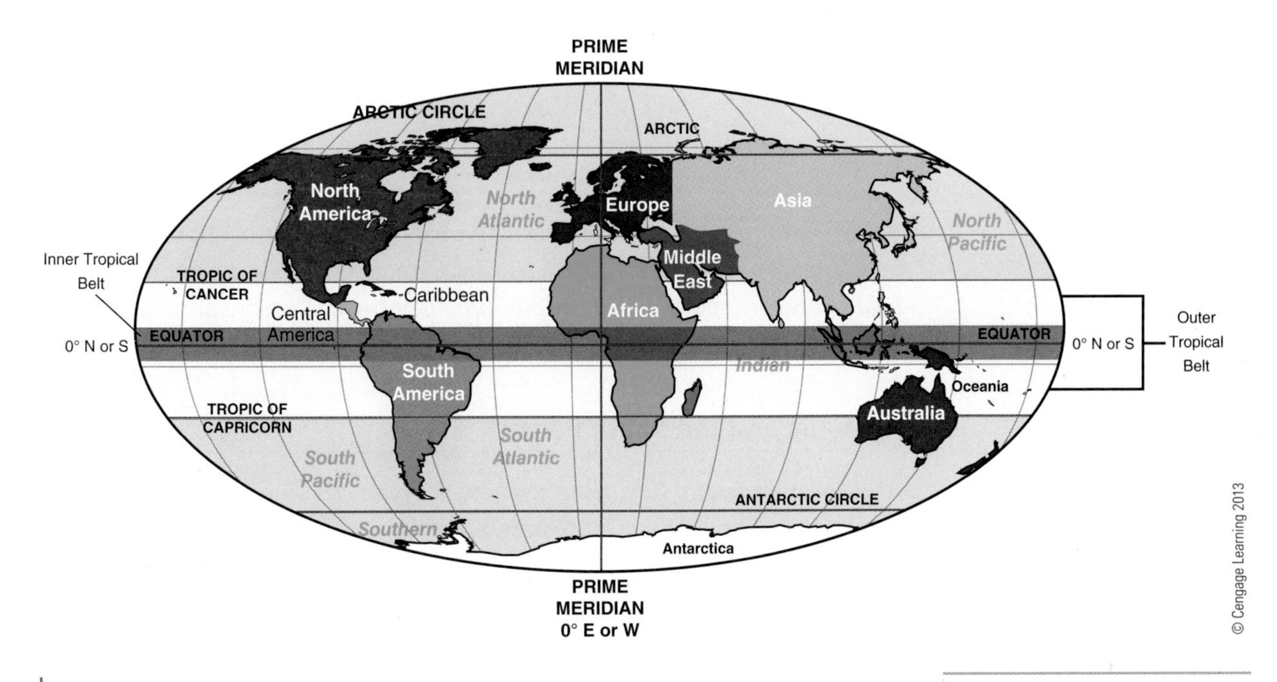

FIGURE 3-9: The inner tropical belt is 3200 miles wide and the home of most indoor plants. It is surrounded by two subtropical belts from 300 to 700 miles wide.

Major Geographic Locales

Most of the plants successfully maintained indoors are concentrated in a belt that lies about 3000 miles north and south of the Equator. The northernmost boundary of this belt is the Tropic of Cancer while the southern boundary is the Tropic of Capricorn (Figure 3-9).

Various changes in the geography within the tropical and subtropical zones along with other climatic differences create much variation in the weather and, accordingly, the plant life. A factor that does remain constant is the uniformity of temperature throughout the year, a factor with wide variation in the world's temperate regions.

Tropical America

Over 112,000 species, 60,000 from the Amazon Basin alone, comprise the known native plants of South America. Many favorite genera can be traced to this region and include *Aphelandra*, *Begonia*, bromeliads, cacti, *Calathea*, *Chamaedorea*, *Cissus*, *Columnea*, *Dieffenbachia*, *Episcia*, ferns, *Fittonia*, *Maranta*, *Nephrolepis*, *Nolina*, orchids, *Peperomia*, *Philodendron*, *Pilea*, *Spathyphyllum*, *Syngonium*, and *Tradescantia*.

Tropical Africa

More than 40,000 known species of plants are indigenous to this continent, such as *Aloe*, *Asparagus*, *Chlorophytum*, *Chrysalidocarpus*, *Coffea*, *Crassula*,

Dracaena, Euphorbia, Ficus lyrata, Kalanchoe, Nephrolepis, Saintpaulia, and *Sansevieria.*

Tropical Asia

Rainy, warm forests in Asia are the native habitats for some of our most durable indoor plants. *Aspidistra, Aucuba,* Citrus, Croton, *Dracaena, Euonymus, Fatsia, Ligustrum, Pittosporum, Podocarpus, Rhapis,* ferns, and orchid discoveries have made their way across the planet.

Australia and the South Pacific

Araucaria, Brassaia, Cordyline, cycads, *Dizygotheca* (*Schefflera*), *Howea* and other palmae, and *Polyscias* are best maintained in well-drained media, similar to that of their native habitat.

Europe

The list of plants from the European continent is short, but graceful ivy (Hedera), well-suited for hanging baskets and topiary training, is native. The tender *Soleirolia,* native to the warm, humid Mediterranean climate, thrives in terrariums.

Our Indoor Comfort Zone

Some people like a cool indoor environment, especially those who are active during work. For instance, diners may feel uncomfortable in restaurants where the wait staff is allowed to adjust the thermostat. As the dinner crowd grows larger, staff members may adjust the temperature downward in order to stay cool while busily waiting on tables. Those sitting to enjoy a meal may have their experience negatively impacted because they are simply too cold. On the other hand, some people are cold-natured and want a warmer room. As the seasons progress to fall and winter, thermostats rise making a space more comfortable and cozy, more like the days of summer.

An optimal indoor temperature is one in which half of the people would like it a bit cooler while the other half would like it a bit warmer, but this of course is theoretical. In the best of cases, a small minority of people would not be comfortable with any given indoor air temperature. Overall, people like indoor temperatures to be maintained between 60° and 80°F. This is quite similar to the temperature range found in tropical forests. Indoor plants are tolerable to all of these temperature conditions as long as air vents are not directed on them. In other words, indoor plants like the same temperatures as people do.

Within indoor spaces, there may be differences in ambient temperatures, the temperature of the surroundings. For instance, the air temperature may be much warmer near heat ducts, near operating equipment such as ovens, or close to lighting or windows with direct sunlight. Interiorscapers need to recognize surroundings with cold temperatures like vestibules in cooler months and air conditioning vents during the summer. In many cases, damage is done to indoor plants from drafts, the movement of air across leaves causing tissues to **dessicate** (Figure 3-10).

FIGURE 3-10: Drafts caused by **HVAC** systems and other sources can be detrimental to plant health.

SUMMARY

Plants and humans can cohabitate indoors because both enjoy similar environments. The object of this relationship is that plants aid in maintaining a beautiful, inviting interior environment when they maintain aesthetic balance. The interiorscaper's goal is not necessarily in the production of new plants, but in the balance of keeping plants healthy and handsome in the spaces occupied by people at work and at home. The provision of an environment similar to the native habitat of the plant's genus will help it to stay better looking, longer.

CHAPTER 4

International Code of Botanical Nomenclature: How It Works

INTRODUCTION

It is quite important to use the precise name of a plant because it provides accuracy of interpretation (Figure 4-1). Scientists must be accurate when working with particular varieties of plants. Informed consumers learn about the appropriate names of plants from many sources including television, Internet outlets, and identification tags provided in pots. It is important for those working with interior plants to know their plants! Recall knowledge enables a professional to work quickly in a business where time is money. It also sets the plant professionals apart in the marketplace because their background knowledge inspires client trust.

Use of Latin and the History of Scientific Naming

Proper identification, classification, and culture of plants allow species to perpetuate. Plant taxonomy, the science of identifying, naming, and classifying plants, provides the organizational structure to link a plant and its name. How does it work?

"Genus" is like a surname, someone's last name.

"Specific Epithet" is similar to a first name.

The system was developed by Swedish naturalist Carl Linnaeus (1707–1778) in the mid-18th century, and it applied to both the animal and plant kingdoms. Prior to his work, plants were identified by long strings of adjectives describing characteristics of the plants. Although a good idea, the previous method was extremely cumbersome. Linnaeus developed a hierarchical system of classification that relies upon increasingly specific categorization of plants. He first devised kingdoms, classes, orders, genera, and species, but later the categories of division, subclass, and family were added. Overall, the system, if graphically produced, would appear to be an inverted pyramid with each level becoming more specific until finally the exact plant is specified (Figure 4-2).

FIGURE 4-1: Plant care professionals feel more confident when they possess knowledge about a wide range of plants.

Plant names are written in Latin, the language of science. Some names have Greek roots, which stands to reason because Greek culture preceded the Romans. Pronunciation of plant names follows Classical Latin guidelines. A guide is provided in the appendix.

The number of binomial names proliferated as new species were established and more categories were formed, and by the late 19th century the nomenclature of many groups of organisms was confusing. International committees in the fields of zoology, botany, bacteriology, and virology have since established rules of organization.

The first word in this binomial system is the **genus** and is a noun. The second word is the **specific epithet**; the word *epithet* means *name* and is an adjective. When typing the names of plants, both the genus and specific epithet should always be italicized and the first letter of the generic name is capitalized. When handwriting plant names, underline both the genus and specific epithet but do not forget to italicize them when using word processing programs.

This system of naming organisms is referred to as **binomial nomenclature**, taken from the Latin roots of bi = two and nomine = name. They are sometimes called **scientific name** or **Latin name.** Both of the aforementioned terms are interchangeable and refer to the same thing.

Take, for instance, the impressive scientific name for Bird of Paradise, *Strelitizia reginae*. The plant was named in honor of plant lover Charlotte Sophia of Mecklenburg-Strelitz, who became regina or queen to George III, King of England from their marriage in 1761 until her death in 1818.

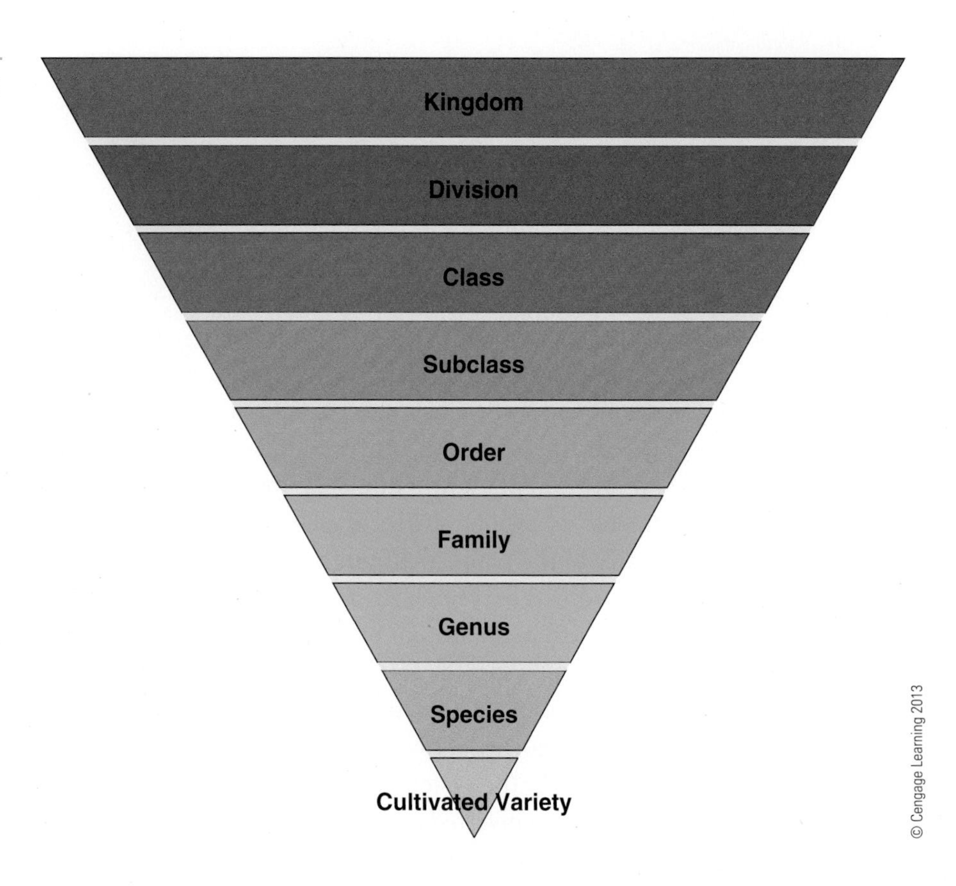

FIGURE 4-2: Each classification level becomes more specific, helping to clarify and identify plants.

When referring to two or more different plants within the same genus, it is appropriate to abbreviate the generic (genus) name. For example, *Dieffenbachia seguine, D. maculata,* and *D. amoena* are all species within the same genus.

Cross-fertilized genera resulting in hybrids are written with an X in front of the genus but are read as "the hybrid genus." For example, a hybrid bromeliad derived from *Guzmania* and *Vriesea* genera is written X *Guzvriesea* and spoken "the hybrid genus *Guzvriesea*."

FIGURE 4-3: Genetic variation through sexual propagation yields various colors of *Viola* flowers.

A third epithet can exist and is the cultivated variety, or "cultivar." Desirable characteristics within a species such as unique leaf variegation or flower color can occur. The differences must be distinct, uniform, and stable in order for these plants to be reproduced. Cultivars are carefully selected, grown by themselves, and are able to be maintained through sexual (seed) or asexual (vegetative) propagation depending on the plant. Cultivars reproduced from seed may show genetic variation; therefore, hybridizers will select only the plants that possess the favorable characteristics and remove those not true to type (Figure 4-3).

Cultivars are written in English or the native tongue of the discoverer and are enclosed within single quotes but are not italicized. A great example of this is the genus *Dracaena*, of which the specific epithet *deremensis* gave rise to the cultivar 'Janet Craig'.

FIGURE 4-4: An Asparagus Fern is not a true fern but a flowering plant. This is a field of *Asparagus densiflorus* 'Myers', Foxtail Fern.

Many plants have the same common name. In some circles Creeping Charlie could refer to about five or six different plants, some with leaves 3 inches across that could fill a hanging basket while another's leaves are only a ½ inch across, a fine-patterned groundcover. The two plants may

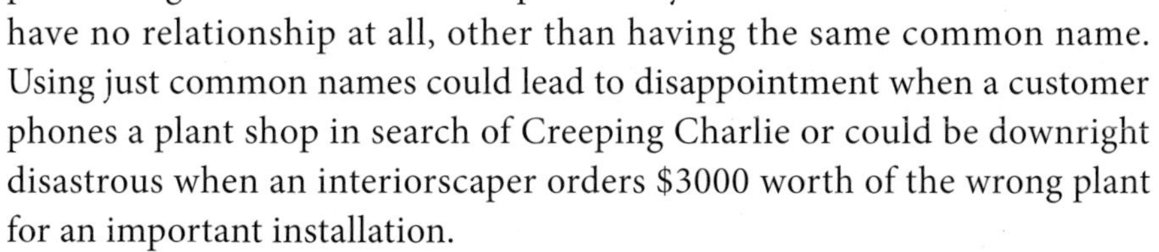

have no relationship at all, other than having the same common name. Using just common names could lead to disappointment when a customer phones a plant shop in search of Creeping Charlie or could be downright disastrous when an interiorscaper orders $3000 worth of the wrong plant for an important installation.

Common names are still present because they can be colorful and, therefore, easy to remember. Take, for instance, the difference between *Sansevieria laurentii* and Mother-in-Law's Tongue. They refer to the same plant, but a tongue-in-cheek joke is easier to remember. Not all mothers-in-law have long tongues, and in this way, common names are often misnomers. We may refer to the ferny and gracefully filmy plant *Asparagus densiflorus* by its common name Asparagus Fern although it is not a true fern. It is in the Asparagus family and is capable of flowering and producing fruit, something a fern will never do (Figure 4-4).

Sometimes, professional horticulturists use just the generic name or just the cultivar. It pays to memorize the names of plants so that when sourcing plants, you will immediately access a file of information about its care, habit, benefits, and even its pitfalls. Of course, the differences between indoor plants due to leaf arrangements, colorations, patterns, and so on help horticulturists to distinguish indoor plants more easily than outdoor, landscape plants. Many people find it easy to recall Latin names because of the adjectival nature of plant taxonomy. When you see the plant, you see the characteristics and are able to recall the name.

FIGURE 4-5: *Equus caballus*.

These explanations may convince budding horticulturists to learn plant names, but there is no need to beat a dead *Equus caballus* (Figure 4-5). Though many sources of plant nomenclature exist, the object is to persevere in use of the most updated names. Ultimately, it is best to follow what your professor or employer dictates. Humility goes a long way.

SCIENTIFIC NAMING CLASSIFICATIONS

Family

A group of one or more genera that share a set of underlying features. Family names end in -aceae, the first letter capitalized and not italicized. Family groupings are broad and their limits are often unclear and up for interpretation.

Genus (plural: Genera)

A group of one or more plants that are capable of breeding together to produce offspring similar to themselves. This is the smallest grouping of plants that have similar characteristics.

Species (singular and plural)

A group of plants with common characteristics, yet different from others within the same genus. They are capable of breeding together to produce offspring similar to themselves.

Subspecies

A distinct variation within a species, between species and variety. Indicated by "subsp." Followed by subspecific epithet in italics.

Varietas (variety) and forma (form)

A variety is a subdivision of a species, but not so distinct as to be called a species. Botanists recommend the use of variety for natural varieties (generally in Latin) while cultivated varieties are termed cultivars (generally not Latin). The term *form* is no longer used, being replaced by cultivar.

Cultivar

A *culti*vated *var*iety with distinct variations of species, subspecies, varieties of forms, or hybrids that are selected or artificially raised. Cultivars are typed in roman type and surrounded by single quotation marks. Sometimes just the genus and the cultivar epithets are used alone when the parentage is not certain.

The following list relates to adjectives used in specific epithets of some indoor plants.

alternifo'lius: With leaves alternately spaced, not opposite.
amazon'icus: Of or from the Amazon River region.
australien'sis: Of or from Australia.
austra'lis: Of or from the Southern Hemisphere.
bic'olor: Two colors.
bifurca'tus: Forked into two segments.
botryoi'des: Resembling the characteristics of grapes.
caeru'leus: Dark blue.
caput-medu'sae: Head of Medusa.
carno'sa: Thick and soft; fleshy.
commuta'tus: Changed or changing.
como'sus: Hairy; with long hair.
crassifo'lius: Thick-leaved.
crena'tus: Scalloped.
crispa'tus: Crisped, curled.
ctenoi'des: Like a comb.
culto'rum: Cultivated, kept in gardens.
dan'icus: From Denmark.
delicio'sus: Delicious, flavorful.

deltoi'des: Triangular, delta-shaped.
densiflo'rus: Densely flowered.
dis'color: Two or more different colors.
domes'ticus: For use indoors.
el'egans: Elegant.
elegantis'simus: Most elegant.
erioste'mon: With wooly stamens.
exalta'tus: Exalted, lofty, profound.
exot'icus: Exotic; from another country.
falca'tus: Curved like a sickle.
fascia'tus: Fanned out.
fejeen'sis: From the Fiji Islands.
flo'ra: Goddess of flowers, festival celebrated on April 28.
fra'grans: Sweetly scented.
ful'gens: To shine.
grandiflo'rus: With large flowers.
helico'nia: From Mount Helicon, Greece; of the muses.
heterophyl'lus: Bearing leaves of more than one form.
hiema'lis: Winter-flowering.
ionan'thus: Violet-colored.
juncifo'lius: Rush-like leaves.
labia'tus: With a lip.
lu'teus: Yellow.
lyra'tus: Lyre-shaped.
manzani'ta: Little apple.
margina'ta: With a margin.
mi'tis: Gentle; without thorns.
nephrol'epis: Kidney-shaped.
ni'dus: Nest.
pal'lida: Chartreuse, pale green, pale.
palma'tum: Palmate.
plumo'sus: Feathery.
podocar'pus: "Foot fruit," bearing fruit on the stalk.
podophyl'lus: With stalked leaves.
pumi'lio: Dwarf.
ra'dians: Radiating; shining.
recurva'tus: Curving back upon itself.
reflex'us: Bent back upon itself.
regi'na: Of the queen.
revol'tus: Rolled back on itself.
rex': King.
scan'dens: Climbing.
scutella'ris: Dish-shaped.
semperflo'rens: Always flowering.

GREEN TIP

Knowledge of scientific names lessens the chances of ordering the wrong plant materials. It is not possible to send plants back to a grower. Incorrect ordering wastes product, time, fuel, and labor.

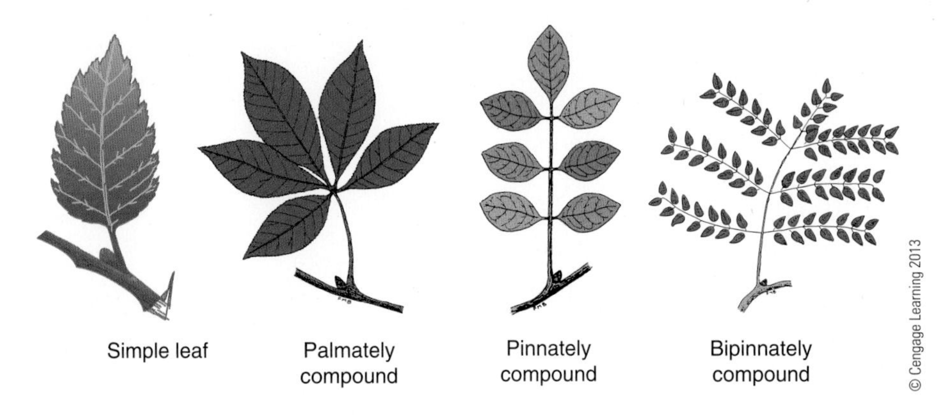

FIGURE 4-6: Simple leaf and different types of compound leaves.

seta'ceus: Bristled.
sinua'tus: Wavy-margined.
spatha'ceus: A spathe surrounding a spadix or flower cluster.
stolonif'era: Producing runners (stolons) that produce roots.
streptocar'pus: With twisted fruit.
termina'lis: Terminal.
trifascia'tus: Three-banded.
variega'tus: Variegated.
zebri'nus: Zebra-striped.

Plants can be identified through observing the characteristics of their forms, patterns, colors, texture, and leaf arrangements; in short, the elements of design applied to the plant. Design elements are classified within the plant world specifically applied to form, arrangement, and coloration of leaves, plant growth habits, and many other details. There is little wonder that they are the fodder for captivating design.

FIGURE 4-7: Parts of a simple leaf. The presence of an axillary bud signifies a true leaf from a leaflet.

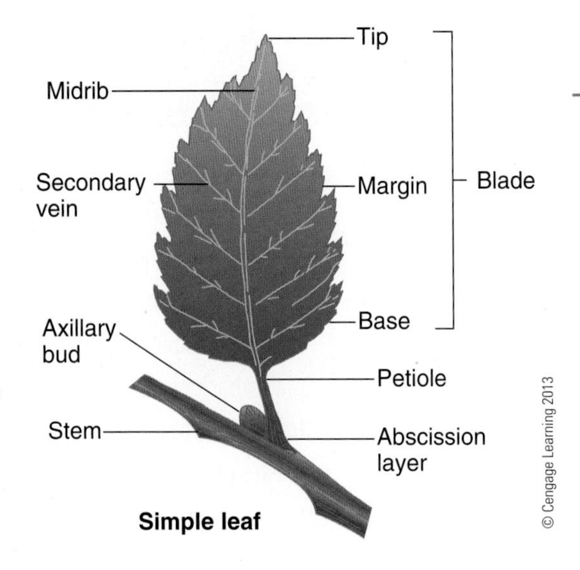

Leaves

Leaves are the food factories of plants and are composed of a thin **blade** attached to stems by a **petiole**. Often, leaf blades are commonly consistent within a group of plants; this aids the horticulturist in identifying a plant. Leaves can be **simple** (a basic leaf consisting of one entire unit) or **compound** (two or more units called **leaflets**; Figure 4-6). Leaflets are distinguished from leaves because only leaves have buds arising from leaf axils, the angle created between the petiole and the stem. Leaflets do not have these axillary buds (Figure 4-7).

FIGURE 4-8: Palmately compound leaves of *Fatsia japonica*.

FIGURE 4-9: *Cycas revoluta* leaves.

FIGURE 4-10: An ostrich feather.

Leaflets are arranged in two basic forms: palmately compound and pinnately compound. Palmately compound leaves have leaflets arising from one point such as fingers arising from a hand (Figure 4-8). Pinnately compound leaves (Figure 4-9) have leaflets attached to a centralized midrib, similar to the arrangement of a feather. The Latin word *pinna* means feather (Figure 4-10).

The way that leaves are arranged on a plant provides further clues for identification through classification. Some plants have leaves placed opposite from each other. Others have alternating arrangements, while others are grouped or whorled (see Figure 4-11). These subtle differences can be quite helpful to identify and distinguish plant materials if noted when learning about a plant that is new to you.

Horticulturists pay attention to the patterns within leaves including the patterns made by the veins, the leaf margins, leaf bases, and leaf tips. A classification of these patterns is made in Figure 4-12.

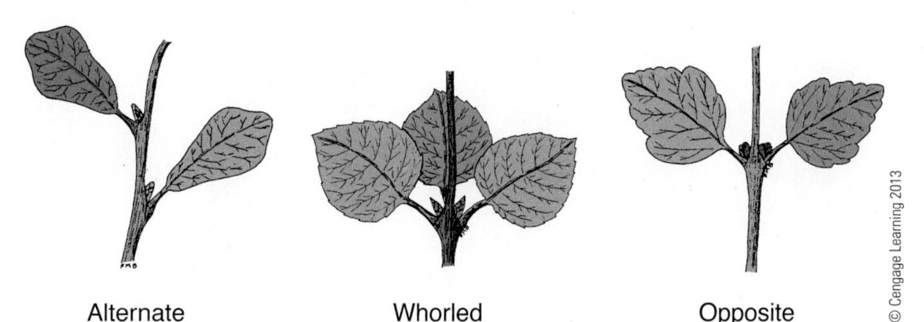

FIGURE 4-11: Alternate, whorled, and opposite leaf types.

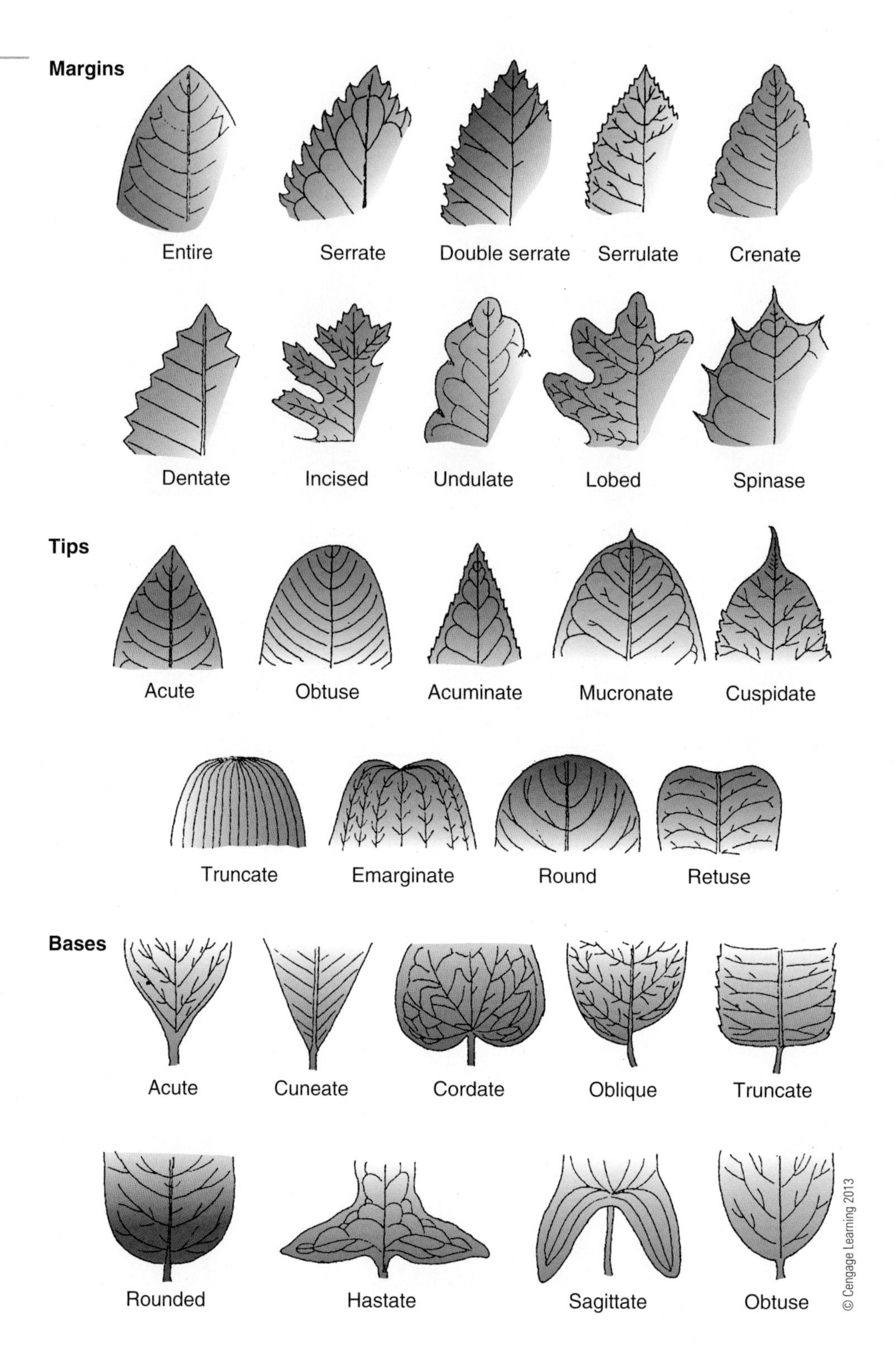

FIGURE 4-12: Leaf margins, tips, and bases.

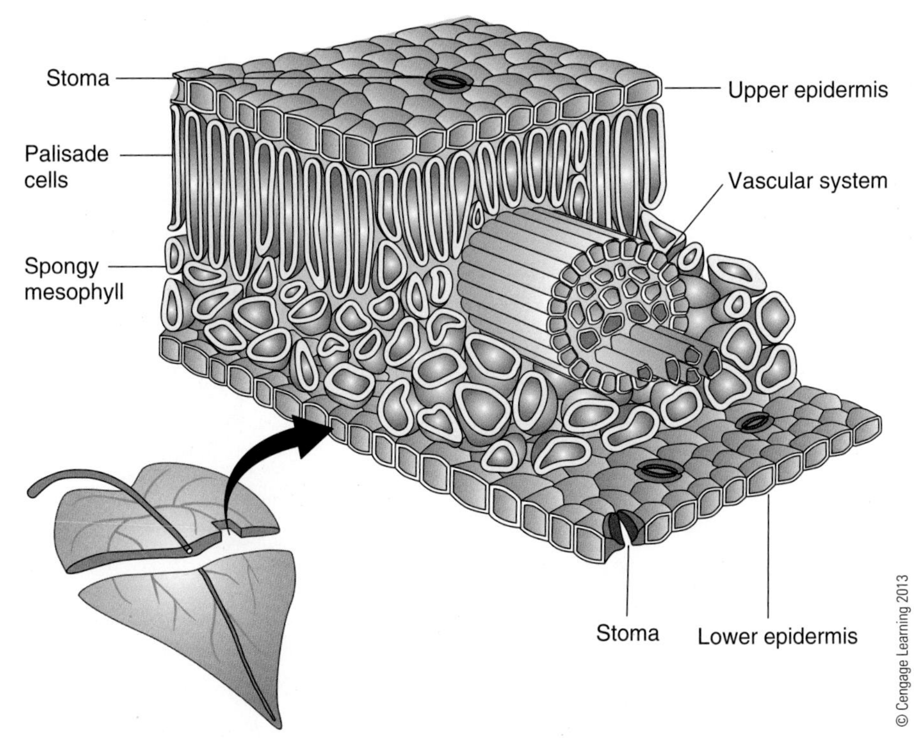

FIGURE 4-13: Cross section of a leaf.

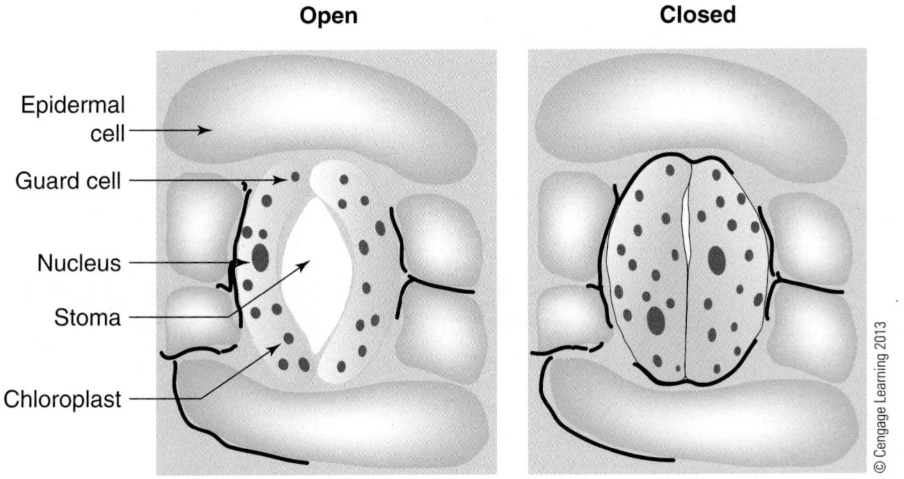

FIGURE 4-14: Stomata in open and closed states.

Although not able to be seen with the naked eye, knowledge of the internal structure of a leaf is important for interiorscapers to understand. The outer epidermis or skin of the leaf consists of a single layer of living cells (see Figure 4-13). Scattered about the epidermis are pairs of guard cells that open and close, collectively making a stoma, a pore allowing for the exchange of gasses from the interior of the leaf to the open air (see Figure 4-14).

FIGURE 4-15: Single *Cyclamen* flowers arise from a fleshy, underground **tuber**.

FIGURE 4-16: *Helianthus* ray flowers are golden, while disc flowers are dark brown in this variety.

FIGURE 4-17: A *Hippeastrum* **scape** arises from an underground **bulb**. This bulb bore three scapes.

Flowers

Indoor lighting conditions are often poor as compared to an outdoor garden or greenhouse in relation to intensity of light energy. It takes high levels of light to get a plant to bloom, so most interior plants that perform well indoors make their way indoors due to interesting foliage rather than spectacular flowers. Since all plants, with the exception of ferns, bloom, the placement and arrangement of flowers, as well as the type of flower, will aid in identifying the plant. For example, flowers may be singular as in a *Cyclamen* (Figure 4-15), composite with multiple ray and disc flowers as in sunflowers (Figure 4-16), or an **inflorescence** with multiple flowers such as an *Amaryllis* (Figure 4-17) or *Cattleya* orchid (Figure 4-18).

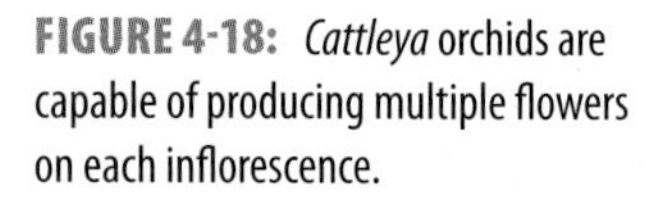

FIGURE 4-18: *Cattleya* orchids are capable of producing multiple flowers on each inflorescence.

Other Characteristics

Flowers and fruit associated with indoor plants may not be present due to lack of light, but beyond the appearance of leaves on the plant, other indicators are present. Look at the aerial stems of the plant growing above the soil line (Figure 4-19). Some stems are herbaceous and green while others are woody. Some stems have patterns from leaf scars while others have lenticels, areas of cells that

FIGURE 4-19: Newer, juvenile growth of this *Ficus* plant is herbaceous while mature growth is woody.

FIGURE 4-20: The stems of *Aspidistra* grow just below the soil surface, so only petioles and blades create the form.

permit gas exchange between stem tissue and the open air. Some plants do not appear to have any stem at all such as *Aspidistra*, where blade and petiole appear above the soil line while the stem remains below (Figure 4-20).

Aerial roots, also called prop roots, provide distinction to plants such as *Ficus maclellandii* 'Alii' (Figure 4-21) and *Phalaenopsis* orchids (Figure 4-22). They utilize these structures for support as well as moisture from the air. Still other plants bear spines as in *Euphorbia*, Crown of Thorns (Figure 4-23).

FIGURE 4-21: A mat of prop roots allows this plant to appear as it would in the rainforest.

FIGURE 4-22: Orchid prop roots swirl about the surface of the pot.

FIGURE 4-23: Thorns are modified stem branches.

SUMMARY

The way plants are named is based upon nouns and adjectives. The binomial system of plant nomenclature uses Latin so that plant designers and scientists can speak the same language in order to communicate accurately. Linking the names to what you see helps you to become an indoor plant specialist. Plant characteristics derived from observing leaves, flowers, and other plant parts, along with growth habits, help to separate plants into more distinctive groups.

Think about the many ways horticulturists view plants, from the tissue level all the way up the chain past the family name. The challenge to the horticulturist is to view plants the way an art historian views a painting: take your time and classify.

CHAPTER 5

Interior Plant History

INTRODUCTION

A basic understanding of the origins of interior plantscaping helps to build appreciation of the rich history of keeping plants indoors. Of course, there has not always been an organized interior plant industry; indeed, it is a 20th-century phenomenon. Throughout history, supporting events and the people who made them happen provide the underpinnings of today's interiorscaping and interior ornamental plant production—important components of the horticulture industry.

Exactly where did this art/science/business begin? As with any other topic in history, vestiges of what we understand interiorscaping to be today are found tens, hundreds, and even thousands of years ago. A common feature found throughout time is that people are drawn to beautiful things. They have an intrinsic need to attain beautiful things, many the products of nature.

Ancient Cultures

Art historians consistently point out the use of plant forms and patterns in art from ancient Egypt, some of the earliest archaeological records available. The first records regarding Egyptian agriculture are about 4000 years old. Prior to this, ancient cultures of Mesopotamia, Babylonia, and Assyria offered innovations in agriculture such as successful irrigation methods. This and other practices were adopted in Egyptian culture. Carvings in stone depicted plant leaves or entire plants in vessels, which displayed cultivation of plants not only for food but also for their beauty. Flowers and plants were grown and presented for pleasing scents, colors, and forms. Flowers and greenery were used as offerings for royalty and divinity alike (Figure 5-1).

FIGURE 5-1: An ancient Egyptian papyrus provides evidence of floral offerings for royalty, as Ankhesenamun brings *Cyperus* foliage and Lotus flowers to Tutankhamen.

FIGURE 5-2: Plant materials were depicted in this Roman fresco from the ancient city of Pompeii, which was destroyed and buried during the eruption of Mount Vesuvius in 79 AD.

The Hanging Gardens of Babylon, one of the Seven Wonders of the World, was a specialized botanical garden of unique and exotic plant materials fed by expert hydro-engineering. The gardens were built by the Babylonian King Nebuchadnezzar II around 600 BC in what is now modern-day Iraq.

In Greek and Roman cultures, people continued to grow plant materials in vessels. The portability of plants in containers allowed for unique plant materials to be displayed as well as traded commercially. Containerized plants allowed for temporary decoration of homes and places of worship.

The Romans utilized decorative plants indoors and in formal gardens at the homes of the wealthy (Figure 5-2). Artwork such as frescoes, mosaic tile images, and sculptures were coupled with vine-covered trellises, potted plants, and flower gardens using myriad genera laid out in symmetrical plantings. Limited use of crystalline mica, a transparent stone, for windows in some villas allowed sunlight to penetrate the interior.

Middle Ages

The first botanical gardens known in history were devoted to growing medicinal plants. Early medicinal gardens were called **physic gardens**, the first of which were formed in 1500s' Europe during the Italian Renaissance. Important plants used for medicine and food were kept throughout the Middle

Ages in monasteries. These perpetuated plant species became more important to society as the feudal system declined and commerce and agriculture increased.

Useful plant materials created interest among nobility and the wealthy in regard to earning lucrative commercial production potential. The ownership of an exotically beautiful, yet unknown, species was intoxicating to those with means to afford such rarities, spurring trade throughout the world.

Renaissance

As interest in exotic plants grew, more funds were directed toward discovery and collections development by states and private individuals in the 17th century. Wealthy European nations established imperial outposts in tropical regions, thus allowing for new plant discoveries and freer flow of exotic plants from native environments to the European continent. Plants showing economic promise due to medical, culinary, or purely aesthetic purposes were cultured in display gardens and buildings, which offered protection during harsh weather.

Today, we recognize the depth and breadth of Dutch floriculture, offering not only the production but inspiring worldwide distribution of floriculture product. Perhaps the greatest early influence on Dutch floriculture was horticulture scientist Charles de l'Ecluse, known as Clusius (1526–1609). Some of his work involved research on the phenomenon of "breaking" of tulips where certain flowers held wildly unique colors and patterns resembling flames and feathering. This peculiarity, later found in the 19th century to have been caused by viruses and not by genetic means, lead to the tulip mania of the 1630s during the Dutch Golden Age. Wealthy merchants paid exorbitant prices for tulip bulbs. Some historians point this period out as being one of the first known economic bubbles, a trade of products with inflated values which, after a period of time, quickly deflate in value (Figure 5-3).

FIGURE 5-3: Photograph of tulips in the Dutch still life art tradition.

Eighteenth Century

Greenhouses as we know them today were not a part of tender plant production in the 1700s. A forerunner of the greenhouse, orangeries, was in place at many royal residences (Figure 5-4). A brick and stone building with large windows, orangeries held heat when temperatures dropped to levels harmful for tender citrus trees valued for their fruit, flowers, and form. During the summer, the plants could be moved outside and placed in various parts of the garden area as accents, kept in decorative containers that today are known as Versailles boxes or Chippendale boxes (Figure 5-5).

Activity and demand for new plant discoveries reached perhaps its greatest heights in the late 18th and 19th centuries. Knowledge about the discovery and care for exotic plants along with their availability, combined with the growing wealth of individuals, created demand for new plants and propagation of existing introductions.

© Snowshill/www.Shutterstock.com

FIGURE 5-4: An 18th-century orangery.

FIGURE 5-5: Garden of Versailles with citrus trees held in Versailles boxes.

© Vladimir Korostyshevskiy /www.Shutterstock.com

Nineteenth Century

Printed materials on the pursuits of gardening were generated in sizeable amounts as demand for information and better methods of gardening rose in the 19th century. Interest in home gardening was considered a wholesome pastime for youngsters and adults. In the United States, the floriculture industry began to build strongholds of production in the northeastern states as nurseries and retail florist shops opened in the cities. The business of horticulture was supported not only by the wealthy but also by the middle class economy, created by the Industrial Revolution, consisting of people with disposable income and the ability to buy floriculture products.

Victorian horticulturists' approaches were broad-based, covering all the information gardeners needed to know, from soil amendments to new cultivars, greenhouse or "stove house" culture, garden layout, and insect pests and diseases, to flower arrangement and suitable color combinations. Publications such as F. W. Burbidge's *Domestic Floriculture* (1874); Paul Henderson's *Practical Floriculture* (1869); Shirley Hibberd's *New and Beautiful Leaved Plants* (1870) and *Rustic Adornments for Homes of Taste and Recreations for Town Folk in the Study and Imitation of Nature* (1856); and *Window Gardening* (1872), edited by H. T. Williams offered suggestions on the culture and display of plants, many of which are still fascinating and beautiful by today's standards.

The indoor garden offered the chance to care and tend to plants without confronting nature's elements. Upscale Victorians showcased their interest in botany by possessing some of the newly introduced plant materials from tropical and subtropical climates. Plants provided the relief of greenery and softened interiors with pattern and texture.

The transport of plants from distant locations was treacherous with journeys on the seas resulting in trips lasting for months with plants subjected to dark stowage and saltwater spray. It was frequently impossible for plants to arrive in good health for propagation, let alone alive. Although it was already known that plants could be cultured in glass jars, British physician and plant enthusiast Nathaniel Bagshaw Ward designed a special transport container, literally a box made with a wooden frame and glass panels, and tested it with a shipment of plants to Australia in the 1830s. The experiment worked, the shipping box proving a successful way of safely keeping small plants alive through the tumultuous seas of Cape Horn.

Victorian entrepreneurs improved upon the idea of these shipping cases, creating "Wardian cases," also called "ferneries" or "terrariums" in various styles and forms, which resembled miniature conservatories. Some fancier

© Dirk Ercken/www.Shutterstock.com

FIGURE 5-6: Terrariums, modern Ward cases, are unique focal points within a space.

GREEN TIP

Consider repurposing a used container for a plant display. A recycled aquarium provides moist soil and humidity to keep ferns and mosses healthy indoors.

types even had built-in water features. Ferneries were able to counteract the somewhat uncomfortable air of the Victorian home environment—drafty, dry, and sooty in the winter months. Terrariums are still a good idea. Circulating water via condensation, they allow for long spans of time between maintenance and can provide ornamental focal areas within an interior design (Figure 5-6).

Ward cases were more of a costly novelty in comparison to other ways plants were displayed. Wirework and string-crafted plant hangers were available for purchase to show off cascading plants. Popular press such as *Godey's Lady's Book* and many others listed different types of plants that could be kept indoors year-round and how they could be artistically displayed (Figure 5-7).

© iStockphoto/Constance McGuire

FIGURE 5-7: A Victorian lady tends to her plants.

Rich-looking palms offered the sweeping lines associated with elegant interiors. *Areca*, *Caryota*, *Chamaedorea*, and *Rhapis* decorated homes, businesses, and special events such as weddings and dances. Many of the plants available in today's marketplace were available by the turn of the 19th century, the forerunners of modern cultivars.

During the 19th century, glass manufacturing aided in providing glazes for greenhouses. The

© iStockphoto/Kjell Brynildsen

FIGURE 5-8: It is still possible to find old greenhouses standing from the late 19th to early 20th centuries.

quality of wavy, bubbled American glass was questioned. Imperfections in the glass could magnify the sun's rays causing burning and scorching of delicate crops within the greenhouse. Higher-quality glass was imported from Europe until long after the U.S. Civil War (Figure 5-8).

Owning one's private greenhouse was considered a status symbol, and wealthy people gathered collections of unusual plants along with able-bodied, knowledgeable staff members to maintain them. Earlier greenhouses depended on the use of centrally located, coal-burning stoves. Vigilant work to keep the stoves burning saved tender plants from succumbing to cold temperature injury and death. Later, the development of steam heat derived from the use of coal and oil fuels to boil water in tanks allowed for better plant maintenance in greenhouses. Steam heat provided higher humidity levels, much more like the plants' tropical native habitats.

Many wealthy patrons had their own conservatories (year-round greenhouses) or winter gardens, sustaining tropical plants and even birds and insects through the cold winter months. This necessitated the employment of staff horticulturists with the ability to keep grounds and greenhouses in order (see Figure 5-9).

© iStockphoto/Linda Steward

FIGURE 5-9: An upper-class Victorian trio enjoys social time in a well-tended estate conservatory.

Larger, urban nurseries began to advertise the placement of ornamental plants indoors by the third quarter of the nineteenth century. Installations could be short-term such as in use for weddings and parties, or plants were offered for outright sale. Society weddings called for the delivery, placement, and tasteful styling of plants in churches and cathedrals. Large palms were banked with smaller plants at the base to create naturalized focal points. Single plants had their pots or root balls covered with swaths of fabric for short-term special events.

There was a keen understanding of the link between exotic plant offerings and their care. On one hand, a growing consumer base valued the luxury of exotic plants for events and long-term displays and learned about their distant origins. Nursery owners realized that people wanted beautiful plants around them but were not necessarily good at keeping them alive. Some retail nurseries advertised the horticultural services of trained employees who could provide necessary care of exotic plants. Through these advertisements, we see the vestiges of the modern interiorscape industry.

Estate owners required some staff horticulturists not only to bring plants from production in the stove house to display in the interior but also to organize the displays. Many horticulture books from this time offered techniques in the arrangement of plants and flowers in the interior, directed toward professional horticulturists as well as homeowners. People realized the need for education and training in order to have a knowledgeable workforce. Information generation and sharing became valuable in order for agriculture to strengthen and prosper.

The importance of agriculture at the national level had been on the minds of legislators for decades. The distribution of information had been going on for years within local agricultural societies and regional associations of farmers who communicated information and techniques to better understand agricultural practices. There was great demand for agricultural research information for families producing food and fiber in the growing nation.

The 19th century gave rise to the Land Grant university system in the United States. Until this time, higher education followed the classical studies model limited to literature, law, mathematics, and ancient languages. Answering the needs for researched, educational information within the industrial revolution, Land Grant institutions were charged with the purpose of teaching agriculture science, engineering, and military tactics. The spirit of this law was the provision of instruction for the common citizen. In 1862, during the height of the American Civil War, President Lincoln signed into law the U.S. Morill Land Grant College Act (Figure 5-10). The U.S. Department of Agriculture was also initiated in 1862. In 1876, the first state agricultural experiment stations were started to harbor agricultural research aiding farmers and industry with the proper methods of agricultural production and management. The Hatch Act of 1887 provided funding for agricultural experiment stations in each state. The need for research information in agricultural sciences was disseminated by faculty to the students

FIGURE 5-10: Land Grant colleges went into effect during the Lincoln administration.

and practicing agriculturists through Land Grant universities all over the nation. Due to the abundant natural resources of the country, institutions and education steadily grew ever since the early years of the Union.

Twentieth Century

With the rise of Land Grant universities, more people were able to take courses and gain degrees in the agricultural sciences. As part of agriculture, horticulture and floriculture gained in popularity because of consumer need for ornamental products to beautify home exteriors and interiors. Flowers and plants were also used to communicate sentiments for funerals, weddings, and special occasions.

In the 1940s, women entered the workforce because men were engaged in military operations during World War II. While many women were occupied in work that would not allow for a homey atmosphere, clerical and managerial positions allowed for decoration of the work environment, hence the touch of bringing houseplants to the office. This nod toward the aesthetics of the workplace, making it tolerable over the weeks and months, is the initiation of the interiorscape industry.

After the war ended, the influx of GIs (members of the U.S. armed forces or their equipment) back to the States caused a major focus in the arts and sciences of the home, unparalleled in American history. Couples built houses, raised families, and furnished their homes inside and out. Varying styles of architecture including the Cape Cod or Ranch style homes along with associated landscape plantings personified the pursuit of the American dream. Such design now provides a historic view of the mid-20th-century domestic landscape.

Interior plants were cultivated and displayed in many homes and businesses. Plant accoutrements were designed to show off plants at their best. Standing floor planters were created from new-age materials such as fiberglass or old standbys like terra cotta and ceramic, and were held in streamlined stands of bent aluminum or wood. Ceramic pots held *Philodendron* plants or blooming African violets on kitchen windowsills. These were not the only plants grown in the 1940s and 1950s. Greenhouses could supply a myriad of different types of plants, and indeed they did. Wholesale and retail greenhouses in North America flourished after World War II to meet consumer demands for floricultural products.

As post-World War II families flourished, children born during that time became members of the Baby Boom generation. As a demographic group, Baby Boomers are people born from 1945 to 1962. Influences from the Korean War and the Vietnam War, and the assassinations of John F. Kennedy, Robert Kennedy, and Martin Luther King, Jr., created social unrest about societal problems and inequalities. Environmental pollution was also a focus of discussion and protest. Some of the industries that had developed from the demands of industrial growth produced toxic waste that contaminated streams, rivers, lakes, air, and soil. Anxiety associated with the lack of environmental quality control came head to head with unrest on university campuses. A focus on the beauty of an unspoiled environment was on the minds of many people. It was during this time of tumult that the Plant Boom of the 1970s was born.

In the 1970s, a popular measure of environmental awareness was obtained by ascertaining the number of houseplants one successfully tended. Plants were set on windowsills, placed in basket-weave plant stands, slung in macramé plant hangers, and displayed in bottles. Terrariums, first used in the 19th century, made a comeback using recycled bottles in clear or green glass. Manufactured terrarium cases were designed in space-age forms for tabletop and floor models in acrylic. Macramé artists pulled out all the stops by tackling double- and triple-tiered plant hangers in acrylic yarns and natural jute highlighted with interwoven beads.

The remainder of the 20th century saw the use of tropical plants indoors, including the use of plants as gifts and for interiorscaping. As industries grew and marketing became more sophisticated, savvy companies understood and used professional interiorscape firms to provide plant products and services. A natural outcome of these services developed at this time—**Christmasscaping**. It became a natural for plant "decorators" to design, install, and take down holiday decorations in malls, office buildings, and other places where their plants were displayed.

Toward the end of the 1900s, many people felt a desire to nest, that is, to furnish their homes with finer accessories and to pay attention to enriched lifestyles and entertaining at home. For years, people had focused more on their careers and less on family and home living. This void created newfound interest in the arts of the home such as interior design and antique collecting, cooking and dining, and gardening.

In the late 1970s, Martha Stewart started a catering company focusing on the entirety of a special event. Fine foods, beverages, and the beauty of floricultural products were highlighted on tabletops. It was suggested that memorable parties and receptions could be held in gardens and other

© kai hecker/www.Shutterstock.com

FIGURE 5-11: Martha Stewart.

natural sites. By 2000, she took her innovative ideas about cooking, interior design, gardening, and other arts worldwide. Through her books, magazine, television appearances, and ultimately her own shows, people learned and gained a newfound appreciation for culinary skill, fine decorative arts, and flowers. She blazed a trail for many other talented designers, media personalities, and networks to illustrate the importance of plants and flowers as elements in entertaining and everyday living (Figure 5-11).

Twenty-First Century

At the start of the 21st century, public and private gardens are important to more people than ever. The end of the 1900s and beginning of the 21st century saw a focus on the importance of great-looking plants in the interior. Escalating energy prices, environmental concerns including global warming, and economic worry have lead to a new focus on environmentalism.

Sustainability issues are at the forefront. There is conscientious conservation of water such as the collection and storage of rainwater. As an example, some botanical gardens have reorganized the way they clean or remove leaves in the fall, allowing beds to accumulate leaves which decay in place, enriching the soil. Forward-thinking administrators and horticulturists realize the psychological links between people and plants, making those connections happen with installations of children's gardens, healing gardens, or themes and events such as butterfly gardens or concerts. Interiorscapes are a key part of this solution, continually providing the natural link to humans indoors.

People are more attuned to plant uses, and plant care information is more plentiful than ever before in our history. Television programs focus on home and garden ideas using ornamental plants. Internet sources provide instantaneous information on plant varieties, origins, care, and design. Marketing efforts expounding the benefits of plants are on the forefront of the interiorscape and floral industries.

As people became more affluent with greater disposable income, the desire for better design grew, as did the accessibility of design. There is a greater appreciation for the beneficial outcomes from people/plant interaction, from aesthetic and therapeutic standpoints. People are learning that plants are much more than seemingly static-yet-living things. The planned placement of plants indoors creates design and has positive psychological and even physical benefits.

Plants are now seen as important, fashionable interior design accents that provide nature's statement indoors. Healthy foliage plants and the occasional spot of flowering plant color speak highly of the residents and associates within such environments. A space without plants lacks life and seems dry and artificial.

Interesting 21st-century plant projects include major settings such as green walls, where entire vertical planes are covered with plants. Imagine three-dimensional wallpaper entirely made from live plants (Figure 5-12)! A small design as simple as a tabletop glass vase filled with **hydrophilic gel** can support a cluster of bromeliads for a long-term display.

FIGURE 5-12: An indoor wall carpeted with tropical plants.

SUMMARY

Over the centuries, people have been fascinated with the display of plants indoors, although they may not have the knowledge and experience of how to care for them. It is interesting to note at the core of this thought people have the need to be surrounded by the beauty of plants. For thousands of years, people have endeavored to foster plant life, which ultimately improved the quality of their lives.

Exotic plants have been sought out from the corners of the planet and tended in exterior and interior gardens. Pursuit of economic interests involving plant species lead to the introduction of fancy-leaved and flowering plants to society. Horticulturists of Imperial Europe cultured plants and recorded data, initializing a knowledge base that aided in the formation of horticultural studies.

The Industrial Revolution brought about a middle class with disposable income, some of which was expended on plants. Plant displays in the 19th-century domestic environment were encouraged and detailed in the popular press as well as in books by horticulturists.

Women's entry into the workforce in post-World War II United States is one of the most significant points in the history of interior plantscaping. Women brought style, comfort, and a refreshed sense of professionalism to corporate environments. Executive decision-makers employed the services of interior designers, who specified the use of flowers and plants to improve surroundings and the company image.

Due to the perishable nature of live plants along with limiting environmental conditions associated with the indoors, corporations found the need to have professional horticultural services to tend to plant needs. The need for plants indoors was strengthened by a back-to-Earth movement in the third quarter of the 20th century, revering all aspects of nature. Architects and interior designers specified plants as important components of new buildings and their interiors. The need for expert placement and maintenance of plants in public places lead to the creation of the interior plantscape industry.

People know that live plants bring a sense of freshness and vitality to living and working spaces. Any room is improved with a healthy, well-displayed plant.

SECTION TWO

DESIGN

It is in people's nature to desire beauty, to surround ourselves with attractive things. We have been programmed for thousands of years to recognize beauty in flowers and plants, the natural splendor around us. Ultimately, the reason why we are bringing plants indoors is to decorate the interior, to enhance the indoor environment with naturally handsome objects. The first goal of the interior plantscaper is to install plants indoors, but this must be done with careful consideration. Of course, horticulturists should be concerned with choosing the right plant for the available light level, amount of care, and other physiological concerns, but a plantscaper is also a designer. Professionals must understand why plants are beautiful and how to showcase them to fully appreciate their attributes. Understanding the principles of design removes much of this mystery enabling the plantscaper to combine various elements and create desirable interiorscapes.

CHAPTER 6

Principles of Interior Planting Design

INTRODUCTION

Understanding the principles of design is akin to unlocking the secret file entitled "What Makes Things Beautiful." Truly, what often separates people with talent from people with knowledge and talent in design-related fields is a thorough understanding of the principles of design. Many designers can generate terrific design whether it is a spin on an existing trend or a completely innovative idea. They have an innate ability to create beautiful design, but they cannot explain it, which may seem frustrating at times.

Another important reason why interiorscapers should memorize and be able to apply design principles is that it is part of our vernacular; it is part of our language. We need to know these terms so that we can communicate with each other in our businesses, schools, and public gardens and with professionals in allied industries. When we speak intelligently, people listen and develop trust in our expertise and ideas. The principles of design can help to bring success (Figure 6-1).

FIGURE 6-1: Learning and speaking about the principles of design help professionals communicate and work toward the goal of creating a beautiful workplace.

The principles of design vary from publication to publication. This text will utilize the following seven principles: balance, proportion, rhythm, scale, unity, dominance, and harmony. It is an objective to memorize these principles, define them, and be able to apply them with several examples.

SEVEN PRINCIPLES OF DESIGN

Because it is important and the first rule of order to memorize these words, a suggested mnemonic, or memory trick, is given:

***B**lue **p**eople **r**egard **s**ummer as an **u**nusually **d**istant **h**oliday.*

The first letters in nearly every word of this sentence stand for a design principle:

Balance

Proportion

Rhythm

Scale

Unity

Dominance

Harmony

It is a ridiculous sentence, but if it helps you to memorize the design principles, it could be quite valuable. Look at it again.

Balance

The first principle of design is balance, which has to do with the distribution of objects in space. People often speak about finding a balance between work and home because they feel that one (usually work) gets more of their attention and time than the other. In order to create a sense of balance in their lives, they need to take away from one side and give it to the other.

Physical Balance and Visual Balance

The notion of weight can be applied to interiorscaping in two ways, those being physical weight, or how much something weighs or its heft, and visual weight, or how heavy something looks. In order to understand this concept, the notion of a central axis, an imaginary line that divides an object or an entire composition in half, must be put into place. A composition in interiorscaping terms could be viewing a leaf, a plant, or an entire installation. It is a view of something (Figure 6-2).

FIGURE 6-2: An imaginary line can be drawn through the center of this *Annas*.

Physical Balance

A plant must maintain physical balance; otherwise, it will topple. Sometimes a plant grows top heavy and requires staking. It is generally considered a fault if an indoor plant must be staked in order to remain upright. Some plants may be top heavy but become stable and physically balanced once the grow pot is placed within a decorative pot. The decorative pot may be heavier than the grow pot/plant unit (Figure 6-3).

Visual Balance

A healthy root system is important for good anchorage. If a plant such as a *Dracaena* cane has a lush foliage crown but an undeveloped root system, it can

FIGURE 6-3: A top-heavy plant with weak visual balance requires staking to maintain physical balance. Although a thin, bamboo stake is somewhat hidden, it is often better when the plant needs no support.

FIGURE 6-4: An underdeveloped root system has caused these *Dracaena* canes to shift during transit.

lose physical balance and lean within the pot creating poor visual balance. The overall plant may be intact, but the unnatural angle looks unstable (Figure 6-4).

Indeed, a plant may have a healthy, vigorous root system that provides stable anchorage for the shoot and leaf portion of the plant above the soil. If the foliage mass is too large for the trunk and pot, the result is a standing tree that appears top heavy. Overall, this plant would be visually unbalanced. The effect of such poor weight distribution negates the aesthetic of a healthy plant and would give clients and their guests an uneasy feeling (Figure 6-5).

FIGURE 6-5: A healthy but top-heavy plant is not aesthetically pleasing.

Symmetry and Asymmetry

Visual weight can be distributed in two different ways within a visually balanced composition—symmetrically or asymmetrically. The best way to view this is by visualizing a playground seesaw. Symmetry by definition is equal weight distributed in the same or similar positions on either side of

FIGURE 6-6: Equal weight distribution on a seesaw demonstrates symmetry.

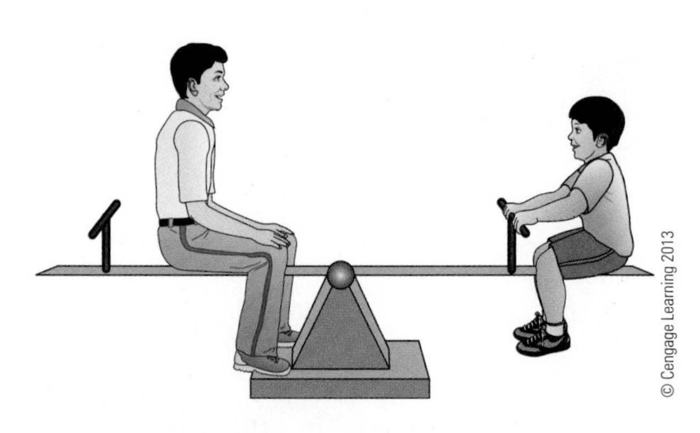

FIGURE 6-8: The larger object must be placed nearer the fulcrum to achieve balance with the smaller object. This is asymmetrical balance.

FIGURE 6-7: Symmetry in plant display.

a static axis. If two fifth graders get on the seesaw, odds are they weigh about the same. They should sit at about the same points on either side of the fulcrum in order to have a good game of balance (Figure 6-6). This scenario applies to the distribution of plants in a setting. A way to bring attention to an entrance is to flank the doorway with identical plants, each the same visual weight as the other (Figure 6-7).

Back at the playground, if an adult plays on a seesaw with a child, the adult must move closer to the center of the board support, or fulcrum. If she does not, the child will undoubtedly become airborne. In this instance, the situation eventually becomes physically and visually balanced, but because one side of the composition has an object that is bigger than the object on the other side, this type of balance is referred to as asymmetrical (Figure 6-8).

Proportion

FIGURE 6-9: The proportions of these two tomatoes differ.

Quantifiable relationships within a given unit can be expressed as proportions (Figure 6-9). Better stated, the bigness or smallness of one thing in comparison to another or the amount of one thing to another is proportion. When we study proportion, we are making a comparison of the amount of one thing to others within a collection or whole.

A good example is found in how people prepare coffee. When some people drink coffee, they love to keep it black to enjoy the richness and bold aroma. Some add milk, so much that it becomes a latte. Further, when some people prepare their own lattes, it seems as though they are drinking milk with a whisper of coffee added to the cup. These proportional choices are

© iStockphoto/rattanapat

FIGURE 6-10: The proportion of these braided *Sansevieria* plants is greater than their pots.

made to suit individual tastes. They are subjective and variable depending on the day and person. Indeed, many people like to switch up proportions to add a different flavor to design.

Being horticulturists rather than baristas, we can relate proportion to the size of a plant's overall mass to its container (Figure 6-10). Most often, the plant's mass is dominant over the container's mass. In this type of proportional selection, the plant is given importance because it is bigger than the decorative pot. This type of proportional expression is the most traditional in plant displays; the abundance of the plant material is highlighted.

This is not always the case in terms of all plant displays. Sometimes, plant mass is subordinate to the container in which it is placed (Figure 6-11). This is a more contemporary approach to plant display. In this way, the container is highlighted over the plant. There may be several reasons for this design approach. One of the reasons why this choice can be made is to emphasize an unusual container, perhaps handmade or specially fabricated for the site. Another reason is that size or height restrictions within the interior must be taken into account. In order to utilize a container with the right elemental statements, the plants must be smaller. Sometimes, too, in order to use a special type of plant, which may seem small, a taller container will bring it to eye-level. The display of plants in tall containers gives them more importance and larger scale.

© iStockphoto/ballycroy

FIGURE 6-11: A seasonal display of *Brassica* and gourds is subordinate in visual mass to the glazed pot.

Remember, proportion deals with the relationship of many elements to each other, not just one to another. A great example here is with the element color and the way colors are combined. Rarely do designers use just two colors in an interior décor palette. Three, five, or even dozens of colors

© iStockphoto/Scovad

FIGURE 6-12: Plants in this apartment occupy a subordinate amount of the vertical space.

can be used successfully. Nature, providing the best examples of color use, paints a sunset with violet, tangerine, and gold. The principle of proportion makes the colors work well together. If you look closely, you will see that one color is allowed to dominate. The allowance of one part of the equation dominating over the others is the key in creating a pleasing composition.

Interiorscaping involves the designed placement of plants to the interior. We relate the size of plants to the space they occupy. As horticulturists, we often want to make a grand statement with plants. We may feel the bigger they are, the better. The Rule of Thirds is handy when building a concept for an interiorscape. Visually divide the height of the room into thirds. Plants may occupy a subordinate or dominant part of the space (Figure 6-12; see Chapter 7, Space, page 101). Some ceilings may be 8 to 10 feet in height, but some are 20 or 40 feet high. This technique can be employed as soon as you walk into a room that is in need of interiorscaping. It gives the designer an instant direction to take in order to make design decisions and suggestions for a client. The same concept can be used for horizontal spaces, such as developing designs for interiorscaping a long hallway (Figure 6-13). A conscious use of this technique may help avoid using plants that are too small for a space, a common problem with inexperienced interiorscapers (Figure 6-14).

FIGURE 6-13: Imagine this space with either a dominant or a subordinate amount of interiorscaping on the horizontal line.

© iStockphoto/xyno

The following example provides a good illustration between proportion and scale. When we look at something being a third of the space, this is an example of proportion. When a designer specifies a plant that is 1/3 the height of the ceiling, and the ceiling is 30 feet high, we now have a scale of measure. We suddenly

© iStockphoto/Scovad

FIGURE 6-14: Quality interior design is not enhanced with plants of very small proportion.

realize that we will source and install a plant that is approximately 10 feet in height. The plant is dominant to human scale, but subordinate to the scale of its new environment.

We can exceed the proportion of plants when creating an interiorscape design. For instance, the Rule of Thirds suggests that plants could be as tall as the ceiling. In some cases, this is most problematic because foliage would be scraping the ceiling, causing a mess at installation and creating the potential for maintenance headaches. Nevertheless, it is an interesting design concept and can be taken a step further. The plant material's proportion can be greater than the imposed limit, suggesting that the plant has grown through the ceiling! You have probably seen this proportion used in specialty designs where the designers wish to impart the feeling of being in a tropical rain forest or taking a walk underneath huge pines. The plants do not actually grow through the ceiling, but are created in a particular form that, prior to installation, appears to have been topped. No matter what, the difference between things like this just happening and their being a conscious choice by the designer is the difference between arranging plants and plant design.

FIGURE 6-15: A pebble wash used as a rhythmic accent within an interiorscape.

Rhythm

A conscious provision of a sense of movement in design is known as rhythm. Plants do move as they grow, their shoots elongate, and leaves turn toward the light. This movement is not detectable to the naked eye, however. A sense of movement makes a design much more interesting and eye-catching than one that seems static or rigid.

Rhythm may be actual or implied. Water moving along within an indoor stream or fountain provides a soothing, focal portion of an interiorscape. It takes planning, materials, and energy to make a water feature happen. A technique that interiorscapers can use that is taken from the natural landscape is to create a **wash**, where pebbles and rocks are laid out in a curvilinear pattern through a large interiorscape. There is no stream present, but water is suggested. Rhythm is implied (Figure 6-15).

A successive line of plants implies rhythm. It allows for a visual pathway through a space. A formal approach is to line up plants, military-style, in neat rows. Such a style suggests a cadence that is crisp, clean, and business-like. The feeling that these words imply suggests that such an interiorscape setting is used for the lobby of a place of business and industry. Some might view it as being static or mechanical (Figure 6-16).

FIGURE 6-16: Ultra-modern, ultra-rigid, the lines of these handcrafted, box-shaped topiaries are repeated by their linear placement.

These designers would break up the straight line and introduce zigzag placements and plants of differing heights. Some people would describe this more broken style as spontaneous and lively, while some might see it as being too random. It is up to the designer to create a statement when it comes to the rhythmic placement of plants; the assertion is most successful when it reflects the environment and philosophy of the client. After all, interiorscapes should be an extension of the client's image.

A conscious use of techniques involving various elements can help the designer impart a more rhythmic interiorscape. The following brief list of design elements provides an opportunity to see more deeply and to apply the things that make up design.

Spacing

When plants are grouped more closely together, they impart a sense of relationship. The brain makes natural comparisons between them. Spacing plants closer together in some areas creates focal emphasis. The space between plants is just as important as the plant itself. Space creates the suggestion of rest and transition. We cannot see the beauty of the plant silhouette without space.

Size

A large plant is naturally more focal in quality. It is sort of like seeing a tall person standing in a room. He is more visible due to his height, so he is easy to see. Why not capitalize on the size of a large plant by making it the focal point in a setting? Use of small plants or those with light visual weight may create weak focal emphasis.

Shape

If you study the form of a plant, it will show you how it is best used. Plants that are tall and thin, or possess one of these attributes, are best used as the skeleton of an interiorscape installation (Figure 6-17). They provide the line that will take the eye into as well as away from the installation. These may be upright plants such as *Dracaena marginata* or plants possessing a hanging line, like *Chlorophytum*. Plants that are massive, possessing the attribute of a rounded shape, provide the muscle for an interiorscape. They provide an anchor to hold attention and draw the viewer's eye into the design.

FIGURE 6-17: This tall Alexander palm forms a skeletal part of the planting, leading the eye to the floors above and the plantings below.

© iStockphoto/Edward ONeil Photography Inc.

FIGURE 6-18: The repetition of white planters and hot pink *Cyclamen* plants creates swift line of rhythm.

Color

Bright color or color contrast is attention getting and attracts the eye. Avoid jerky color rhythm. Underplantings of all the same or similar color *Guzmania* provide a smooth flow of color rather than underplantings of many different colors. A more professional look is given when colors relate through repetition (Figure 6-18).

Scale

At the heart of the principle of scale is *measurement*. As interiorscapers, we should communicate a particular size plant within a measured space. It is important that interiorscapers communicate in measurable means. Often, we hear someone stating that a plant is low light or that a plantscape does not need much water. What do these things mean? Better clarity is established when we tell a customer that a low light plant stays healthy when the office lights are left on for 12 hours or that you should only water succulent plants when no particles adhere to your fingertips. Providing scale in communicating and installing interior plants keeps the project objective. New, smart designers should carry tape measures when working on site specifications. A handy thing to do is to measure the length of your foot, your stride, and the distance from your elbow to fingertips. Knowing these lengths provides designers/interiorscapers with a constant measuring scale.

GREEN TIP

Using the same style and color of planter in an interiorscape not only stresses the design principle of unity but also can be more earth-friendly. When companies package planters for shipping, they often offer them in units of 3, 6, 10, or 12. If they are packaged together, less cardboard and plastic material is required for safe shipping.

Unity

Given the human propensity of bringing the beauty of nature indoors, there is a union of the rugged outdoors with the artificial world of the interior. Plants provide a sense of unity with the outdoors and make people feel more in touch with nature. Unity on its own is a celebration of similarities. There is a strong statement when design elements are unified. When people are united, there is a feeling of focus toward accomplishment. It is satisfying.

We can think of unity when it comes to groups of people, united for a cause or a concern. Initially, those people are drawn together because they have much in common. They have the ability to make an organized assertion. Sometimes, the designer's approach can be overstated and the principle of unity can help to objectify this problem. Installations can be overly unified, causing a look that seems too repetitive or just-out-of-box. Using all the same genus of plants or all the same form of plants or containers can cause this problem. It can look boring or unstudied. A better approach might be to use all the same color planters, but vary their form. This idea adds a sense of harmony to the design.

Dominance

In nature, there are aspects of living things that are naturally dominant. For instance, it is unusual to see two plants growing at exactly the same height. Some plants dominate others due to their capacity at maturity. They may have large leaves or be vigorous growers. Dominance may be seen in plants that have bright, advancing colors such as Croton or flowering plants versus *Schefflera* or *Aglaonema* (Figure 6-19).

In discussing dominance as a design principle, we can find similar terms in focal point, or focal area, and emphasis (Figure 6-20). Focal areas, a term that is better than the term focal point because it is more of an area than a

FIGURE 6-19: Differences in height, color, or pattern can make one species stand out from the others.

© photoinnovation/www.Shutterstock.com

specific point, by necessity should be the most dominant part of a composition or a design. A focal area is the part of the design with the highest concentration of elements. Take, for example, an aerial view of an indoor atrium. A central fountain is flanked by color crop (blooming plants) and concentric rings of foliage plants. Walkways radiate from the water feature and individual beds are planted with a variety of plant materials. In this setting, the fountain is the focal area due to its form, concentration of closely spaced plants, and their reflective colors.

Focal areas bring emphasis to a portion of the design. They help the eyes and brain focus on what should be the first thing a viewer sees in a design. The eye will naturally distinguish the focal area at first, if it is well designed. From the focal area, the eye should be able to easily follow lines within the design, yet just as easily return back to the focal area. Emphasis allows a designer to create eye-catching interest within a design or within an environment.

FIGURE 6-20: The tip of a permanent botanical Christmas tree in this photo is an example of a focal area. The spun layers of plastic form a tight cluster, accented by the brown strands.

© iStockphoto/Kurt Drubbel

Plants can lend dominance to architectural features, making them focal. Designers frequently use the technique of using plants to frame a special entrance, lobby, or elevator doors. This technique lessens confusion of the client's guests. This technique is especially useful when there are many doors or hallways present for people to find. Entrances that are more important can be plantscaped to bring them attention, aiding in traffic flow. This is why an interiorscape designer should spend some time at the potential client's site, preferably making a visit well before the initial sales call. Similarly, two- or three-dimensional works of art gain more attention and are more important when flanked by symmetrically or asymmetrically balanced plantings. Taking time to observe how spaces are used or how foot traffic flows creates a more informed designer who can ultimately make great design suggestions and increase company revenue. It is also a good idea to allow designers who have never seen existing accounts to make a visit. Perhaps they have new ideas or have learned of innovations that can add to and improve an existing interiorscape.

Naturally dominant plant materials should be used in a way to highlight their features as well as the entire plant. Individual plants or plantings can be made more prevalent with specialized lighting or by featuring the plant alone, surrounded by space. For instance, special plants are more dominant when they are isolated, such as a *Podocarpus gracilior* with its lacy foliage and graceful, arching branches. A healthy specimen of this plant would look great if incorporated with other plants, but the subtle beauty of its form, texture, and pattern might be hidden. On its own, without clutter, the client could recognize its beauty and higher value. Such isolation emphasizes individual plants (Figure 6-21).

FIGURE 6-21: The Ficus tree in this lobby is dominant because it is surrounded by open, negative space.

©NDI--Natural Decorations, INC.

FIGURE 6-22: *Oncidium* orchids and Maidenhair fern, made by hand. The sum is greater than the parts of this magnificent display of permanent botanicals.

Harmony

Where unity is a celebration of similarities, harmony celebrates differences. A restaurant offering only peanut butter and jelly sandwiches daily, with no substitutions, would soon meet its demise. A diverse palette of plants can sometimes be the ticket to a successful installation. Designers need to take care not to offer too many different kinds of plant materials without some repetition of elements in the same areas. It is important to note that harmony is expressed when the whole product is greater than the sum of its parts.

Consider a combination of artificial plants as an interior accent (Figure 6-22). Permanent plants are made from simple elements: fabric, wire, and paper. In the designer's hands, they have become contemporary statements giving the *feeling* of tropical plants, with no care.

Another approach to arriving at a clear theme is the use of preserved plant material combined with live plants (Figure 6-23). Massive plants could be difficult to replace once an installation was completed and business was operational, such as a 24-hour coffee shop in a casino. Many designers elect to install plants made from preserved plant materials. The fact that they are made from natural plant products coupled with their distance from viewers and, in some restaurants and clubs, low-intensity lighting, provides a *trompe l'oeil*, supported by underplantings of live plant materials.

© iStockphoto/wang cheng

FIGURE 6-23: Plants at eye-level are living while taller, high-value plants are permanent botanicals made from organic bark and fabric foliage. Initial investment is compensated over a period of time because the permanent plants would only need cleaning, not replacement, and plant servicing is relatively fast.

Think about some of the restaurants that use many kinds of antique signage to portray a theme. The signs they use advertised soda, detergent, or hair dye. What some deem trash, others are able to collect and display with artful, delightful effects. The ability to create harmony through eclecticism is important for designers in any medium. Mixed inventories of containers, odds and ends, plants left over from previous installations, and variations in ground covers might create design havoc or, in the hands of a skilled designer, could result in a masterpiece.

Harmony is learned over time through practice, reading, and observation. In other words, designers bring not just their immediate skills but also their lifetimes of observation, knowledge, and practice to the table.

Design is Personal

Designers often take fierce pride in their work, and this is a great attribute. Creating an interiorscape design depends on the designer bringing forth all of his past knowledge and experience. A project may appear to take a few days or a few months to develop, but in actuality, it has taken many years to bring to fruition keeping in mind classroom and on-the-job learning.

It is important for designers to remember that there are many stakeholders who can also be decision-makers in developing a plantscape. It pays to be thick-skinned during the development of a design and installation. It is the interiorscaper's job to bring the ideas of many other people to fruition. A wise philosopher once said, "Do not fall in love with your own work." Always leave room for change because if you feel your work is perfect, there is no room for improvement.

Of course, all the people associated with an interiorscape company are affected by these concepts of design; therefore, they should have an appreciation for the creation of a beautiful environment. As you read this text, you should be able to see the information from many different professional angles. Some will take this information and become plantscape technicians, aiding in the installation and maintenance of plants in the interior. The work of the technician is perhaps the most important work in the plantscape company. They are the faces and voices of the horticulture company. They are also the caregivers to the plants on display. Their knowledge and implementation of good plant-care practices can make or break the company's bottom line. Much of the company's success rests on their capabilities. Many people start their industry careers as hort technicians. Such a background provides them with a world of knowledge that can aid them in better understanding the company's practices. These types of positions generally have less responsibility, especially outside of the shift assignment. When the workday is finished, the employee can leave and generally does not have to think about his job. Similarly, the pay for this position reflects the responsibility level. Working for an interior plantscape company as a technician, whether part time or full time, is a great job for those in or just out of school. It is also a mobile skill. If a partner must move to another location, a trained horticultural technician can take his skills with him, find a good job, or develop his own horticultural enterprise. With a good attitude and a nose for opportunity, people with horticultural knowledge and skills can find more work than they can handle.

Some people reading this text will work in management of horticultural providers. They must realize that the company's most important investment is in their employees. They may be the personality associated with initial plant installations or solving larger horticultural problems such as pest control or nutrition issues, but often their chief job is managing people. They make sure that plant care rotations are staffed and that shifts are covered when someone is ill or on leave of absence or if an employee fails to show up for work. It is a juggling game where one has to take care not to make too many mistakes.

Some who learn the contents of this text may be involved in allied areas of horticulture. They may enter careers in nursery management, floral management, garden center management, or many other areas. They may find employment with companies that do not have interiorscape departments. Even if this is the case, some companies could explore the possibility of initiating such a department, in which case an employee with a course in IP would be valuable (Figure 6-24).

FIGURE 6-24: Interiorscape departments arise from existing garden centers, nurseries, or florists.

It is not uncommon for students to use components of information they learn in an interior plants course and apply their knowledge in many ways. For instance, it would inspire confidence when working in a retail floral department or store to know the names of many plants along with their display or growth requirements and presentation techniques. Knowledgeable horticulturists do not shy away from plants but gladly accept the opportunity to work with them.

Some students get discouraged in IP courses because they have difficulty spelling the Latin names of plants. Some have problems studying and memorizing. Frustration can occur if a student is good at the science of horticulture but not as good with design aspects. It is not uncommon for some to create fantastic interiorscape design plans but make poor plant selections. There is room for all types of talent in interior plantscaping. We hope that readers will not be discouraged and not expect to be immediate masters in IP work. It takes many years of on-the-job training to be good in anything, including interiorscaping. It is best to learn from your mistakes as well as the mistakes of others. Taking a course on interior plantscaping is in many ways an orientation for professional work. Most of the people with whom you will work did not have the opportunity to take such a course but would have loved to do so. It is important to appreciate and respect those who have taken risks and learned their work on the job. Befriend them, if possible, for they have much valuable information and skill (Figure 6-25).

FIGURE 6-25: A passion for plants, a will to work, and time investment builds the best horticulturists.

SUMMARY

The principles of design define various aspects of what makes an interiorscape beautiful to behold. By taking time to learn them, horticulturists unlock the secrets to appreciating plants indoors for both the client and the employee. All stakeholders should be enlightened to good design before, during, and after an installation.

Design relies upon creativity. There are many good solutions for an interiorscape need, so a certain freedom of creativity should be a part of the interiorscaper's makeup. In the same way that a blank canvas offers many ways it can be painted in color and pattern, so too an interior environment offers many opportunities for successful planting design.

CHAPTER 7

Elements of Interior Planting Design

INTRODUCTION

The elements of design are the things that make a design. A well-made cookie consists of butter, sugar, eggs, and perhaps chocolate chunks or nuts. The materials are blended at the right proportions to create a delicious taste, and the dough is formed into a dollop so that it will bake in a somewhat symmetrically rounded shape (Figure 7-1).

FIGURE 7-1: Cookies are made up of different elements.

Sometimes cookies are baked for a special occasion such as a wedding or holiday. Sometimes, they are made for everyday enjoyment. The same can be said for combining various ingredients to create an interiorscape. Once a client makes initial contact, plantscapers begin the creative process and define needs for height, form, width, and color combinations. These and other elements are the ingredients of interiorscape design.

The elements of design are broad and transcend all types of design, from fashion to aeronautics to horticulture. Within each of these industries, millions of combinations of elements can be made; thus, no two designs can truly be said to be exactly the same. Even when we try to replicate a plant design, nature's infinite variations cause each plantscape to be different.

Great plantscape design is born from an artful combination of design elements rather than repetitive use of the same elements. In order for plantscapes to sing, designers have to expand their knowledge of the ins and outs of design elements (Figure 7-2). In other words, play!

FIGURE 7-2: Interesting round and spiky combinations of lighting, furniture, and plants interact and create exciting design.

Color

It is important for interior plantscaping students to have a running knowledge of what color is and how we can work with it. The science of color tells us that all colors are absorbed into an object with the exception of the color that is reflected, which then is the color that we see.

It is best to quick-sketch a 12-slice color wheel. Know that there are an infinite number of colors; therefore, there is no definite stopping point between any colors. It is traditional to draw the color wheel with separation lines, but in reality they do not exist. Anyone who pays attention to nature knows that color is something to be explored and appreciated. The primary colors are red, yellow, and blue. The secondary colors are derived by combining two primaries; thus, red and yellow create orange, yellow and blue yield green, and blue and red provide violet. There are three primary colors and three secondary colors (Figure 7-3).

Tertiary colors are developed from the combination of one primary with one secondary color.

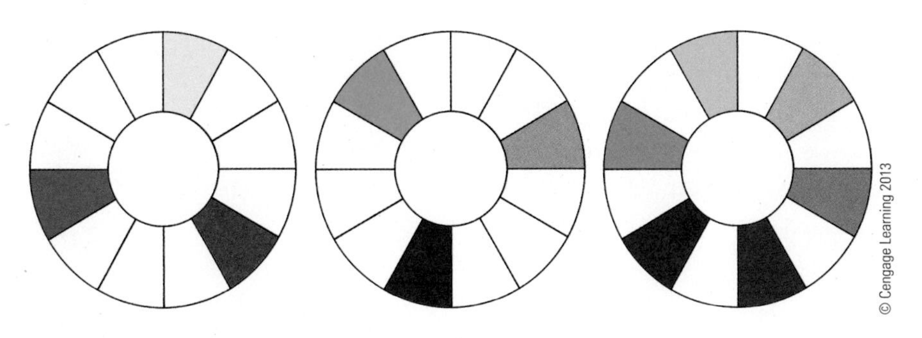

FIGURE 7-3: Primary, secondary, and tertiary colors. The primary colors are foundational from which all other colors are derived. The combination of two primary colors in various proportions creates a secondary color. Tertiary colors are a blend of a primary and an adjacent secondary color in various proportions.

Red–orange
Yellow–orange
Blue–violet
Red–violet
Blue–green
Yellow–green

Looking at the color wheel, you can see the arrangement of the colors in terms of the way they blend.

The color wheel has two regions of temperature (Figure 7-4). One side, consisting of blues and greens, provides the perception of lower temperatures; it feels cool. The opposite side gives the feeling of warmth. This color psychology is immensely important because it sends powerful messages to anyone who views the colors. These colors also suggest emotions. The cooler side of the wheel is calming because it is reminiscent of the colors seen in the ocean, lakes, sky, and glaciers. Hot colors suggest sunlight, fire, and molten lava. Many clients would appreciate being reminded about color temperature

FIGURE 7-4: Color temperature can evoke a feeling of coolness and calm or warmth and activity.

FIGURE 7-5: Deepening shades of emerald green are seen in this landscape.

response. It makes sense to use greater proportions of calming colors in waiting rooms as opposed to advancing colors.

Understanding Color Terminology

Many different terms are used to describe the quality of colors. Knowing the subtle meanings of color terms can aid in communicating with other design professionals. Colors that are derived from natural light, such as the colors seen from the rainbow made with a prism in sunlight, are **hues**. A hue can be manipulated as in making paint colors. If white is added to a hue, it becomes a **tint**. If black is added to a hue, it becomes a **shade**. Tints and shades change a color by either lightening it or making it darker, thus changing the **value** of the hue (Figure 7-5). If a combination of white and black is added to a hue, the hue becomes grayed and is known as a **tone**. Adding gray to a hue lessens its **intensity**. A highly intense color has little to no gray in it.

Horticultural Neutrals

Since green is most often the color used in greatest amounts, it may or may not be a part of the planned color scheme depending on the effect the designer wishes to achieve. Other neutral colors are browns, tans, whites, gray variations, and blacks because they naturally occur among plants. The paler the color, the more neutral it becomes. Highly tinted colors, highly shaded colors, and grayed-down, tonal colors blend very well with other colors.

Combining Colors

As interiorscapers, we work constantly with variations of the color green in live plants. There are seemingly millions of different tints, tones, saturations, and more. Understanding and appreciating the subtle values of green in chosen plants has ramifications on how they are used. Perhaps the veins in a plant contrast with the shaded part of the leaf blade. This may set a color palette for associated pots and accessories. The amount of available light, whether artificial, natural, or a combination of both, has a major effect on color. For areas that are dim, sometimes the best selections are surfaces that will reflect even the smallest amount of light, that being white, light tints of colors, and shiny, reflective surfaces.

Having a working knowledge of color combinations is handy for the interiorscaper and, for that matter, a necessity for anyone involved with design. When discussing colors with clients or members of allied trades/industries, it is far better for the horticulturist to highlight the positive aspects of color specifications rather than dwell upon elements that seemingly will not work. Keep in mind the aspects of color intensity and temperature to build the right mood and theme for the design in following the choices of interior designers and decision-makers.

Take a few moments to identify the color or colors used in the greatest proportions in your current space. Identify other colors and assign percentages to all colors used.

In this respect, remember that all colors do work together! Nature shows us every day that obscure or seemingly distasteful colors can be combined to make uniquely beautiful design. The keys to working with color are proportion, theme, available light, and sales skill.

Whenever hues are combined, whether that means two colors that are very similar to each other and low in contrast all the way through the combination of many, highly contrasting colors, the proportion of colors used should vary. Allow one color to dominate within a setting to achieve success. In working with interior plants, most often values of green will dominate a setting. Although this is the case, any other color used in an installation, such as that in variegated foliage, planters, or flowers, will stand out because it is the thing in the design that is *different*. Considering that plants are mostly accents in interior design, odds are the color of the wall covering or carpet may dictate the dominant color of the space.

A sunset, a cityscape, a walk in the woods, or the colors used in masterful paintings can provide valuable information for combining colors. Study objects or settings that are beautiful to you and observe the number of different colors used. After identifying the colors, place an approximate percentage on each of the colors. Some percentages may be close, but nearly always there is dominance of some colors over others (Figure 7-6).

FIGURE 7-6: Homegrown, organic tomatoes with a greater proportion of advancing reds and golds, and then cool green sepals.

FIGURE 7-7: A monochromatic color scheme involves a hue and any of its tints, tones, or shades.

FIGURE 7-8: A close-up view of a custom-built planter in monochromatic red.

Monochromatic

A mono (one) chromatic (color) scheme consists of one color from the color wheel. The color can be manipulated by adding white, gray, or black to it. For example, red paint mixed with black provides maroon paint. In a monochromatic plantscape, red tulip plants could be installed with pink hyacinth in cranberry-colored planters for a short-term, springtime plant display. Monochromatic color combinations are highly unified because all colors used are derived from a single base color. All the colors seem to be in agreement. It is always best whenever possible to use a wide value of the base color for interest. In our previous example, a top-dressing of dark red stones would add a fourth value of red to the installation (Figures 7-7 and 7-8).

Complementary

A complementary combination relies on two colors that are direct opposites on the color wheel. This combination provides the strongest contrast between two different colors. This combination is more dynamic than monochromatic. In other words, opposites do attract. This combination can be quite attractive whether it is done boldly with pure hues (*Crossandra* in blue Delft porcelain) or subtly with tints (pale blue *Sinningia* in salmon pots). These examples provide interiorscapes in miniature, but it is also admirable to think big on a project. The same color effects can be derived on a larger scale.

People often feel that a red and green complementary color combination is going to look Christmas-y, but once a 'Janet Craig' is dropped into a candy-apple red cache pot, any further combination will pale in comparison. As in any blend of colors, always allow one to dominate over the others to provide a strong, professional design statement (Figure 7-9).

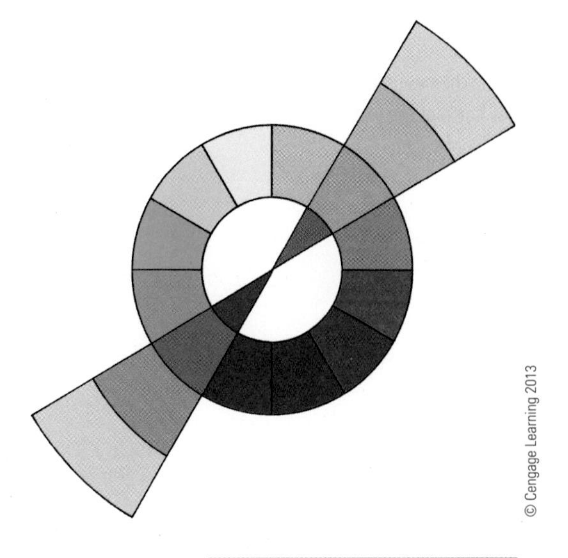

FIGURE 7-9: Complementary color schemes use opposites on the color wheel and may include tints, tones, or shades.

Analogous

An analogous color combination selects two to four *adjacent* colors on the 12-spoke color wheel. This is often a popular combination of colors because it is not repetitive, nor is it too contrasting to most people. Imagine an interiorscape setting with *Croton* 'Norma', red-centered *Neoregelia,* and citrus orange *Guzmania*. This tropical display would be the center of attention because they are all advancing colors and would show up well in even low light conditions. This combination would certainly bring life to a dull corner or to an area that needed to draw attention, such as below important signage a reception area. These bright colors would get the attention they long for, and from a marketing standpoint, whenever people notice plants they are more inclined to want them, which is good business for the entire horticulture industry (see Figures 7-10 and 7-11).

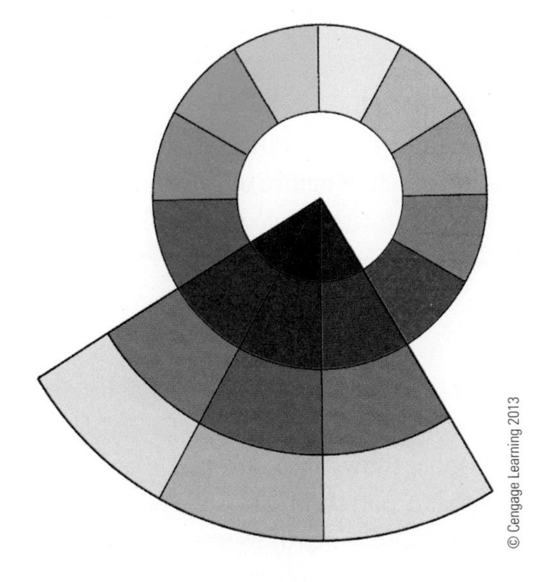

FIGURE 7-10: Analogous color schemes are created from two to no more than four flanking colors on the 12-spoke color wheel.

FIGURE 7-11: Intensities of red, red-violet, violet, and blue-violet are represented in this mass of horticultural bounty.

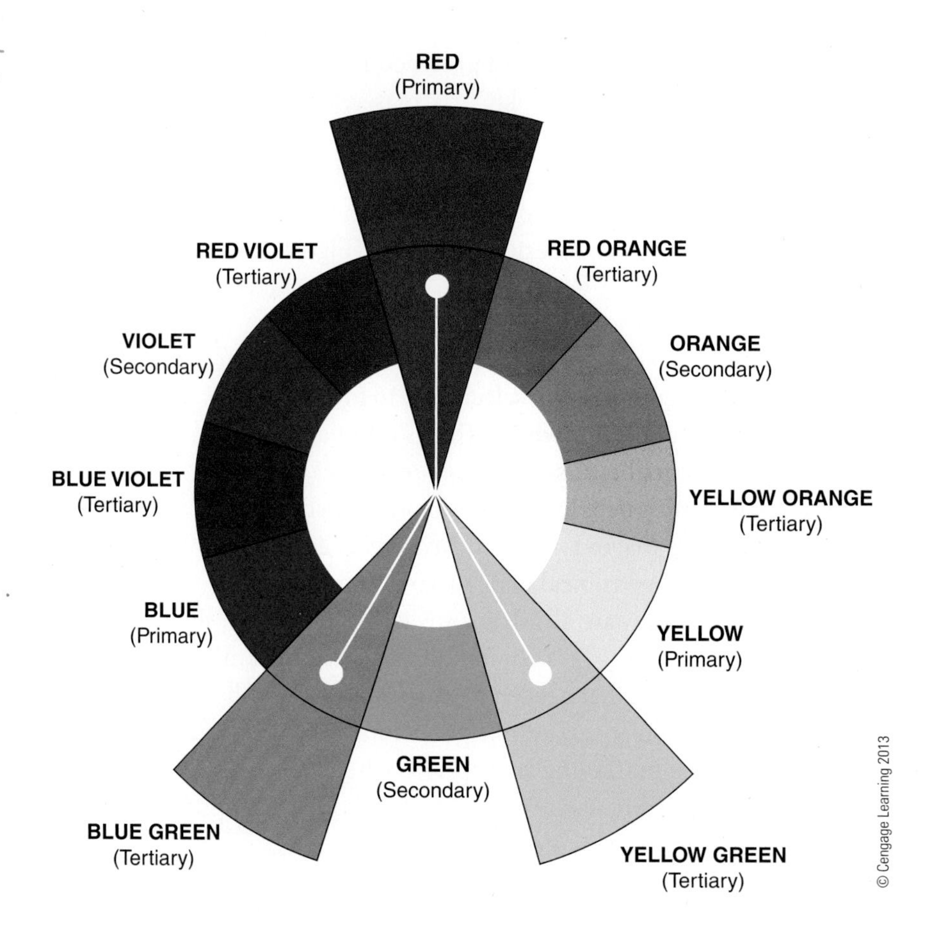

FIGURE 7-12: Split-complementary color schemes utilize three colors—any hue plus the two hues adjacent to its complement.

Split-Complement

In creating a split-complementary scheme, a hue is chosen, then two hues on either side of the direct complement are chosen. For instance, blue-green can be partnered with orange and red. The nature of this combination places differing colors with semi-strong contrasts to maximize color impact (Figures 7-12 and 7-13).

FIGURE 7-13: Blue-violet, red-violet, and yellow *Freesia* form a split-complementary mix.

Polychromatic

Sometimes, a more riotous color effect is desired, perhaps with the installation of a temporary indoor garden setting. Many interiorscapers and other horticulturists install displays for seasonal shows, store promotions, or any time a highly colorful planting is needed. A polychromatic scheme may successfully employ tints, pure hues, shades, and tones of four or more colors. Again, dominance is achieved by the skillful use of colors in varying proportions, with consideration to available light, theme, and client expectation. It is important to meet the objectives of the client and other stakeholders as closely as possible (Figure 7-14).

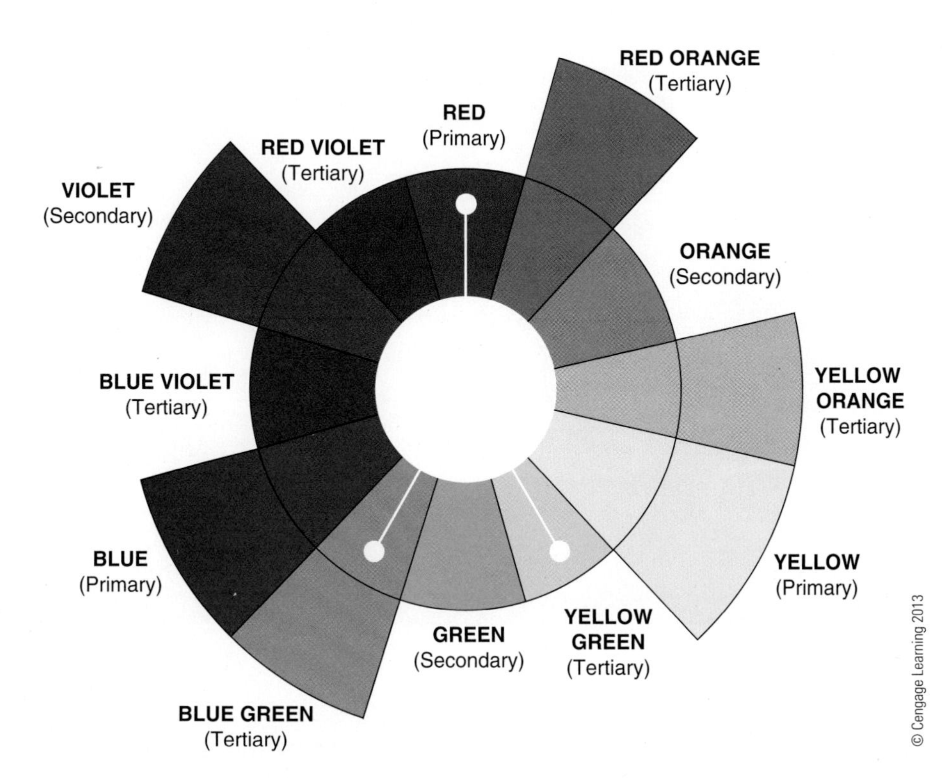

FIGURE 7-14: Polychromatic color schemes use many different colors.

Color Psychology

Color has an impact on people triggering feelings and emotional responses from the subtle to the dramatic. Much of this has to do with the evolution of humankind within the environment. For example, it is rare to find many foods that are blue-green or black. Our minds tell us that if we see foods in these unusual colors, it is a signal that they may be spoiled or otherwise unhealthy to eat (Figure 7-15).

We work with a lot of different colors when plant foliage, flowers, and containers are considered. The colorful as well as not-so-colorful environments of our clients usually provide the starting point for color added by our plants, pots, and other accessories. The world of the interiorscaper is artistic, yielding influence and feeling.

Black

Black is a color of mystery to many people. Like the night, black hides things. It is a very fashionable color considering the appeal of black cocktail dresses and that many designers like to wear black. Perhaps they like the color because it causes more of a focus on form and less on the color itself. A black jardinière can add depth and dramatic flair to a room and make an ordinary plant seem more sophisticated.

FIGURE 7-15: Blue tomatoes seem artificial, yet amusing. Would you eat this?

© iStockphoto/Robert Rushton

FIGURE 7-16: Different values of blue jardinières would allow the interiorscaper to use this elusive color indoors.

Blue

We associate calm water and skies with blue. People who are true blue are good for their word, trustworthy. Blue is on the cool side of the color wheel. Some people feel that too much blue is cold and lifeless while others feel cooler in otherwise warm spaces because of it. Consider this color for serenity and tranquility (Figure 7-16).

Brown

Soil, earth, leather, chocolate—brown conjures feelings associated with natural products. It is considered to be a masculine color but may be balanced with more feminine colors in tints or pure hues for impact (Figure 7-17).

FIGURE 7-17: Green accents lend life to earthy browns.

© iStockphoto/Alexander Tregubov

Green

Green is abundant throughout nature in places where vegetation is lush. It is a refreshing color because it reminds people of ocean water, grass, and the forest. Celebrities discuss waiting in the "green room" before their appearances on television talk shows. Designers use greens in abundance for these and similar settings to help people feel calm and relaxed (Figure 7-18).

© Stephen Coburn/www.Shutterstock.com

FIGURE 7-18: Green life is necessary for indoor spaces, including the kitchen.

Orange

Orange, like red, stimulates the brain and heart. It makes people feel active, hungry, and wanting to move about. It is a warm color and is interpreted as cheerful as well as pushy (Figure 7-19).

Purple

Purple, a mixture of red and blue, can have greater proportions of one of the parent hues or the other. If greater in blue, it appears cooler in temperature, recedes, and can take on a more peaceful feeling. If greater on the red side, it is warm and is more readily seen. Some consider this color pensive and regal (Figure 7-20).

FIGURE 7-19: Gray-green *Tillandsia* on gnarled branches provide cool relief to color and pattern activity.

© iStockphoto/Chris Gramly3

Red

Red gets attention. It quickens the heartbeat and respiration rate of those exposed to concentrated areas of the color. It is used for stop signs and the apple in the Garden of Eden. Red plant materials are often associated with Christmas

FIGURE 7-20: A wall of windows and light colored floors and ceiling allowed the interior designer to specify a more receding, fashionable color such as lavender for an office hallway.

and Valentine's Day. Red is used as an accent more than the dominant color in most, but not all, environments (Figure 7-21).

White

Surfaces and environments in the color of white show every bit of dust and dirt. It could be a challenge to use white planters on the floor where they could be marred and scuffed. Nevertheless, the crispness of this color is undeniable with the bright, white spathe blooms of *Spathiphyllum* and the way it is chosen for clean, pristine themes (Figure 7-22).

Yellow

Sunshine and springtime flowers are characterized by the color yellow. It is a color signifying caution, from the yellow of street signs to the alternation with black in stinging yellow jackets. It is a vibrant, vibrating color (Figure 7-23).

FIGURE 7-21: Pillows and vases provide transition from the pure hue red carpet.

FIGURE 7-22: Even a luxury condominium would be sterile without live plants.

FIGURE 7-23: A live plant would calm this bright yellow hallway.

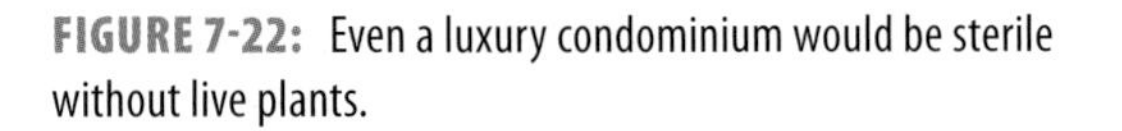

Pattern

Interiorscapers work with plant materials exhibiting scores of beautiful patterns. Pattern, a repeated visual image, is seen in the repetition of pinnae (singular pinna, a palm leaflet) in the leaf of a *Howea fosterana* or *Rhapis excelsa*. Such patterns are a treat to view and enjoy, and can be further appreciated when illuminated. Patterns will be cast upon walls and ceilings creating additional positive visual effects.

Lighting from below elongates a design adding a sense of height, greater scale, command, or even foreboding. Lighting directed from straight above an object may tend to visually flatten the object, adding emphasis to the upper leaves.

Generally, directional lighting, where light is directed from angles, is best because it seems more dynamic or changing like the angle of the sun. Consideration must be given to the light fixture placement itself, which must be safely out of the pathway of foot traffic. It must also be innocuous and work well with the interior design.

It is generally best to use a contrast of patterns in a given setting. Environments that already possess a high level of patterning such as intricate wallpaper designs, walls covered in small-scale stones, or other highly patterned backgrounds would benefit from the use of plain, smooth foliage plants such as *Ficus lyrata* or *Ficus elastica*. A designer might choose these plants because they possess broad, smooth leaves without variegation or intricate patterns. They would stand out in comparison to their background (Figure 7-24).

Conversely, a loft space with minimal interior decoration would find a striking focal area with the use of a *Howea fosterana* or bamboo (Figure 7-25).

GREEN TIP

Decorative pots and planters that have lost appeal due to scratches and wear can be revitalized with faux finishing techniques. Various techniques of paint application result in a finished planter that looks better than its original state and is customized for the client's interior.

FIGURE 7-24: Designers specified plain chairs and matching planter to lend calm contrast to large patterned wallpaper.

Consideration must also be given to the feeling or theme derived by plant patterns. Palms may give the feeling of a tropical retreat, but they may also suggest Victorian society spaces or even modern, urban style. A *Ficus* tree is versatile and has been used frequently to replicate the feeling from temperate zone trees such as maples or oaks, although it has no relation to the two. The fact that the leaves are elliptical and that there are so many of them aids in their being of vague origin once they are teamed with *Hiemalis* begonias, for example, as an underplanting. Such a combination would work nicely for a garden-themed restaurant.

Understanding the contrast of pattern provides the interiorscape designer with more possibilities for design rather than fewer. Some designers feel as though they only work with the same 5 to 10 species of plants, hardly ever varying from them because they are durable and can withstand poor conditions. With depth of knowledge of the many kinds of interior plants available on the market along with a working knowledge of the principles of design, greater variation allows for designer variation. Of course, practice is of the utmost importance. Just because a particular plant may be suited for the environmental conditions in a specific space does not mean that the designer must be inelastic to change. Wholesale and retail costs, accessibility for watering and cleaning,

FIGURE 7-25: Two works of art, a *Ficus* tree and a sculpture, possess much pattern yet are subordinate to white space in this loft condominium.

sourcing, and many more challenges must be addressed. Good interiorscaping is the result of about 10% designing and 90% management issues.

Form

It seems like a bad thing to lump all the fantastic forms of plants together into a few categories. A large part of the fascination with indoor plants is the high level of variation in their forms, the growth habit in which the individual leaves and stems take to provide an overall shape to the plant. We can combine materials into groups that have similar form characteristics to help with design decisions. This then becomes a technique to aid designers in choosing the right plants for a given space. It is not the only tool they need, but it is quite helpful in making design decisions.

Plants can be classified into the following categories (Figure 7-26):

Linear
Hanging
Upright

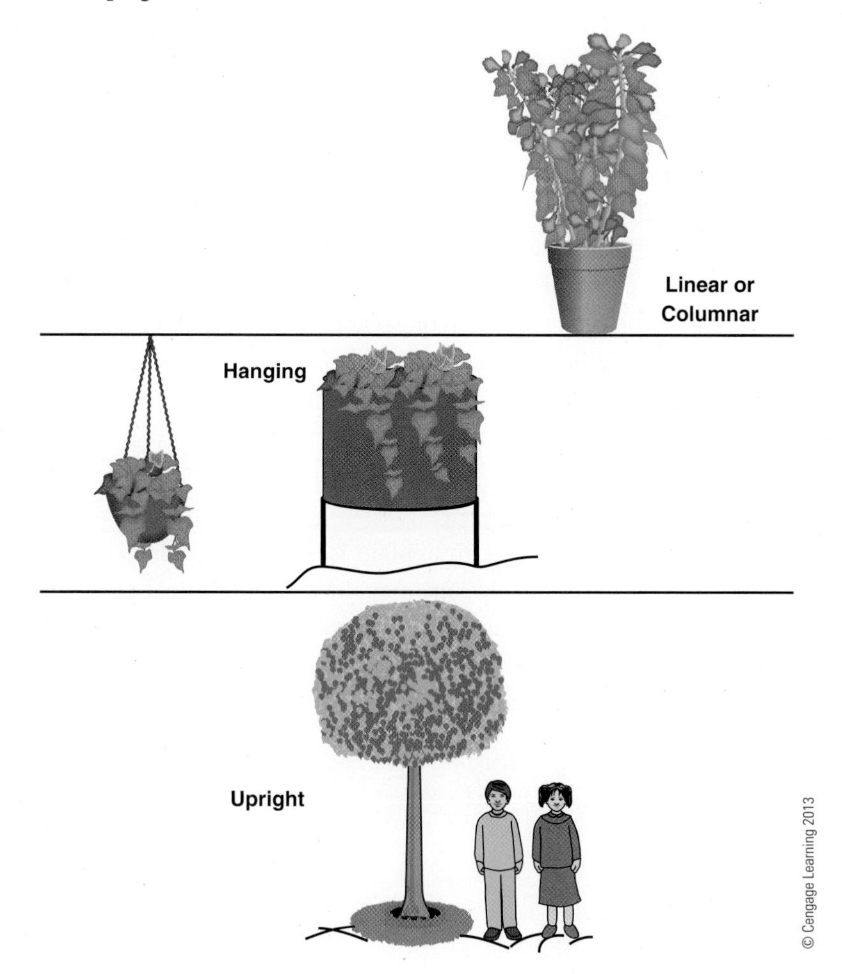

FIGURE 7-26: Categories of indoor plants are most often characterized by being linear or columnar, hanging or upright, with bare trunk and foliage canopy.

FIGURE 7-27: A columnar form of canes topped with foliage makes this *Yucca* well suited to frame a door or window or soften an architectural column.

Linear

Plants in the linear category are line-dominant (Figure 7-27). It is possible to see a column form to the overall plant. These plants can form the skeleton of an installation with different forms or look attractive by themselves. Some plants in this category naturally have attached foliage from the rim of the pot to the stem terminals. Some plants such as Dragon Palm may have exposed stems with bursts of linear leaves at terminals or may have foliage from top to bottom. The growth points for plants can be at floor level or suspended and pendulous. By keeping this in mind, designers will use plants in imaginative ways. Some examples in the category include:

Balfour Aralia
Bamboo
Chamaedorea erumpens
Dracaena marginata
Euphorbia candelabra
Ficus sp.
Schefflera elegantissima

Mass

Mass plants are generally full and rounded. They provide muscularity to an interiorscape and look fine by themselves or when combined with linear plants for variety in form. This combining can be accomplished with great effect with plants of different form placed together in one planting or in separate pots, but grouped together. Unifying elements such as pots of the same color and form could provide a sense of interaction and agreement (Figure 7-28).

FIGURE 7-28: Overall, this *Spathiphyllum* is oval and full in form.

Accent/Filler

Sometimes, the addition of a third form of plant is necessary to draw a relationship between forms. Lush installations may make use of accent plants to fill an overall pattern. They also create softer transitions between plants or plants and planters. Accent plants may be upright, spreading, or cascading (Figure 7-29) and include the following:

Ficus pumila
Hedera helix
Philodendron scandens
Scindapsus aureus
Setcreasea pallida
Tradescantia zebrina

Distinctive

These plants are sculptural and require space around them so that their unusual shapes can be admired. For instance, the versatile, tough *Dracaena marginata* fits not only into the linear category, but some specimens would also be considered distinctive. Such plants would have been grown and trained to achieve spreading, twisting stems with pom-pom-like bursts of foliage or perhaps continually pruned to grow solid, contorted trunks.

Many orchids and bromeliads look best when surrounded by plenty of space. In this way, their assertive bloom spikes do not entangle or become obscured by other vegetation. It is possible to mass these plants together for color impact or for sheer luxury. The downside of this technique is that the humble, lighter form, the **wabi-sabi** of the plant, is lost (Figure 7-30). The following list of plants can be classified as distinctive and, as such, used with plenty of negative space around them so their unique forms can be more easily seen.

Phalaenopsis spp.
Guzmania lingulata
Phoenix roebelenii
Bonsai plant materials
Oncidium spp.
Yucca spp.

© vnlit/www.Shutterstock.com

FIGURE 7-29: Pothos is often used as a surface underplanting to fill space around larger plants.

Note that the plants listed previously provide examples that fit into categories of form, but many could fall into varying categories. Scale has much to do with the use of plants in certain categories. A plant that may be a mass plant in one setting may become a filler/transition plant in a large-scale planting.

When combining plants in a design, a very simple technique is to use three heights in one planting. Select a plant that is tall, one that is intermediate,

© Walter Pall/www.Shutterstock.com

FIGURE 7-30: It has taken years of careful culture to nurture this *Crataegus* into its current form. It requires space so that its beauty may be observed.

FIGURE 7-31: Many Gesneriads have velvety leaves. Gloxinias also have velvety, bell-shaped flowers and grow from underground **tubers**.

and a third that is short or cascading. This high/medium/low technique of selection is very helpful in many design applications. It can be applied to using three heights of plants in a single planter or clustering three plants in three separate jardinières. Keep this technique in mind for making combinations in dish gardens, Christmascaping designs, and placement of decorative accessories.

Texture

Texture refers to surfaces of objects. Texture can make pattern, but pattern does not make texture because texture appeals to the sense of touch. It is not necessary for most people to reach out and grab a handful of prickly *Opuntia*. The very act of it could require hours of tweezing to remove the miniscule quills! Nevertheless, this sort of thing is done by children who have not had the unpleasant experience of tangling with a cactus. Once a person has been through it, he or she generally does not seek it out again.

The sense of feel is stored in our memory bank, and for interiorscapers this is good thing. Horticulturists usually do not invite people to feel plants unless it is a direct object of the planting as in the case of gardens for children or the visually impaired. Constant handling of tropical plants can result in mechanical damage such as torn leaves, necrotic areas, and more. Passers-by wanting to sense whether a plant is live or artificial, over time, will damage seemingly durable faux plants.

Leaves of plants possess texture. Often, tropical plants have a low-grade sheen that is a product of their cuticle layers. Some plants have leaves covered in minute hairs that give them a velvety appearance. Light is not reflected from them but sinks into the leaf (Figure 7-31). Veins and mid-ribs provide additional features to a leaf's texture. Textural differences can be a part of the proportional play with plant placement. By repeating some of the plants throughout a planting, their texture will also be repeated thus allowing for dominance of a texture.

FIGURE 7-32: This Sago palm, its leaves, stem, Spanish moss, and pot all have varying textures, which adds character to the selection.

Decorative pots as design elements are made with varying surfaces. The surface of a planter can be analogous or contrast to the texture of the plants' leaves (Figure 7-32).

Space

Designers do not always consider space as a consciously selected element because we tend to think of things like plants as either being full and healthy or weak and spindly. Looking deeper into a design and the relationship of plants to each other and their surroundings brings about the realization that different kinds of space are necessary to see. By using space, it is possible to enjoy things on a deeper level.

There are two types of space—positive and negative. Positive space is an actual item, be it a leaf, a plant, or an installation of plants. Negative space refers to open space, air, or a void. It is necessary to have negative space around an object in order to more clearly see its pattern.

An object may be seen as being simple or beautiful because of the fact that it is surrounded by space. Important religious icons are separated from other objects in order to help establish their importance. Sculptures are placed on stands or plinths in order to elevate them to higher spaces and make them more visible. The same can be accomplished with establishing the importance of plants, dish gardens, hanging baskets, and many more horticultural offerings. The use of dominant negative space is linked to Asian culture, for instance, Ikebana, Japanese floral arrangement, and bonsai, the physical dwarfing of plant materials for aesthetic purposes. Such art forms utilize space around individual placements in a design or the removal of buds and branches to establish space between branches (Figure 7-33).

It is a technique of design to lessen negative space between objects, drawing them closer together to achieve impact. For example, rather than placing several single-specimen plants about an entire office building floor, a more focal approach would be to draw them together, creating a more unified statement. This stronger cluster creates more visual impact versus lessening the

FIGURE 7-33: Careful pruning establishes more aesthetic growth patterns and negative space to view them.

statement by increasing the distance between plants. Of course, this is only a plant placement technique. Many factors associated with an individual account should be analyzed and respected in order to achieve client satisfaction.

Negative space usage in any type of design is important to consider and should not be equated with budget constraints. Negative space in music is called a "rest." A similar technique is used in speech where the speaker slows down his or her speed. Such pauses cause people to listen more closely. A contrast in space can cause viewers to pay closer attention to an interiorscape planting.

Line

Designers approach line in their work in terms of line direction. At a basic level, plants grow upward from the ground level; many designers approach their work this way, without variation. The study of individual plants' growth habits, both natural and trained, provides a wealth of opportunity for placement and design. This is one of the reasons why it is important to learn about many different plants as a horticulturist and see them as elements of design rather than being good with one alone. Consider the fact that many durable interior plants do not have lignified stems and are able to cascade as they grow. Pothos is grown as a tabletop specimen, in a hanging basket, or on a pole, armature, or frame. This is a tough plant in its own right. These different ways of growing and retailing the plant are quite brilliant and creative. There are probably 100 other types of indoor plants that could be adapted to different growth forms, all variations on line direction. The exploration and development of such products is up to the reader.

Line Direction

The flow of a line provides determination on the way we perceive the line. The eye follows lines in a distinct direction, and the brain interprets the line's pathway to mean something.

Vertical Line

The vertical line suggests strength and growth. It is the most common line direction used in interiorscaping due to the growth habit of plants. It is an energetic line because it takes energy for things to grow or rise upward. Designers refer to it as the line of duty.

Horizontal Line

Horizontal lines are restful and placid. Consider a calm lake and the dominant, horizontal line of the water. When people and animals rest, they recline horizontally to sleep. A horizontal line of plant materials appears balanced.

Diagonal Line

A diagonal line is the line of change. It is seen as possibly rising or falling. When a plant such as a Miniature Schefflera grows on a diagonal line, it may appear to be lacking in visual balance. Some indoor plant enthusiasts may try to stake the plant in order to gain visual balance. By doing this, the plant may appear to be more of a contraption than a thing of beauty from nature.

A more pleasing example of a diagonal line is seen in a planting of Creeping Fig, *Ficus pumila*, trained on a brick wall within an atrium or conservatory. The blank wall becomes much more interesting with a dynamic line of the plant growing across it. The line established by the plant's growth

HOW LINE INSPIRED A NEW PRODUCT

In the early 1980s, a woman who was also a wife and mother started an interiorscape business from scratch. Barbara Helfman tried to figure out ways of increasing sales of plants, gaining new clients, and perhaps selling more plants and maintenance contracts to existing clients (Figure 7-34). She thought about how many of her commercial clients provided their employees not with individual offices but with more economical and efficient cubicles. The cubicles worked well to provide noise reduction, giving employees a sense of organized space, but they appeared sterile and impersonal.

Barbara thought that if she could install plants not necessarily within the small cubicle spaces, but on the top edge of cubicle walls, they could be shared and enjoyed among several employees. Her "partition mount planters" would provide live plants, bringing life to the seemingly boring colors and feel of large expanses of cubicle space. These simple but exceptional planters sold like hotcakes, and nearly every account she had ordered more. Since that time, thousands of interiorscapers have bought Topsider planters and installed them in for their clients, proving that a great idea is worth the trouble of development.

© Nursery Retailer

FIGURE 7-34: Barbara Helfman, inventor and businesswoman. An example of how product development arises from fulfilling industry needs.

would suggest a garden under change, where plants were allowed to meander and grow on their own without constant human tending. Transcending the horticultural aspects, the overall feeling that could be derived from this planting is more relaxed. This is a space where nature is allowed to take its course.

Hanging Line

Hanging lines are languid, suggesting a steamy, tropical jungle, a tangle of plant growth, or a cascading waterfall. Plants that have stems and growth habits adaptable to such displays are highlighted in hanging-type plantings where they are held in suspended containers or wall-mounted receptacles. Container supply companies offer plant holders that can accommodate suspended plants, but designers of interiorscapes should not feel as though they can only work within the realm of what is available at nearby stores. If you can design with plants, you are a great candidate to design planter holders as well. Part of the creativity of being a designer includes stretching your work into other areas of design that have an effect in your area.

Using suspended plants is a good idea if potential problems can be worked out ahead of installation. Suspended plants are a way of using plants in a space when floor space is at a minimum. The environment can have all the benefits of a plant display without the plant being in someone's way. Indeed, the plant would be out of harm's way from someone knocking it over or moving it, which could be detrimental to the display life of the plant.

SUMMARY

Once elements of interiorscape come together, magnificent and original combinations can occur. Designers need to be dreamers, conjuring up combinations of colors, textures, and forms. They should challenge themselves to avoid repeating elemental combinations that have been previously successful. When we reject design elements, for example, stating, "Hanging plants are just too much trouble," we eliminate the possibility of creating something unique. Every challenge thrown at us should be met with a positive embrace. A new plant material, an unusual color request, or a dream of using plants in novel and unimagined ways should be at the heart of every interiorscape company.

CHAPTER 8

Complementary Product and Design Services

INTRODUCTION

The world of products and services in which interiorscapers can specialize is immense. Good interiorscapers may think of their companies as distillers of the beautiful world of horticulture for interior spaces. We do not have to limit ourselves only to the installation and maintenance of indoor plants, although many companies successfully do so. Indeed, the forms, combinations, and care for live plants can keep a company busy in perpetuity; however, our clients have additional needs for interior horticulture design. With interest and will, interiorscapers can offer additional products and associated services. After all, we gain entry to the places where people work and live. This, along with horticultural knowledge and abilities, makes us the best candidates to supplement traditional, live-plant interiorscaping with silk plants, holiday displays, and floral designs (Figure 8-1).

FIGURE 8-1: Interiorscapers can provide permanent botanical displays, some of which are premade and ready for installation.

FIGURE 8-2: Realism in plant replication is appreciated by designers and clients. This example copies *Sedum*, *Echeveria*, and other plants in an artful combination.

Permanent Botanicals

Permanent flowers and plants offer interiorscapers creative possibilities for the use of alternative plant forms indoors. Decades of design and development now offer the marketplace a myriad of plants and flowers, from the ultrarealistic to the ultrafantastic, all created with the idea of providing designers with new media for plantscaping in an imitation of nature.

For some horticulturists, the very thought of using an artificial plant is more than they can bear. Trained and practiced for years in the propagation, production, display, and culture of ornamental plants, some plant people may feel that artificial materials are not worthy of use.

This viewpoint is limiting for horticulturists who appreciate the form, patterns, and overall beauty of permanent plant materials. In today's market, they are no longer called artificial but are referred to as *permanent botanicals*, *faux botanicals*, *silk plants*, or *silks*. Permanent botanicals can provide a world of design inspiration adding elements of design that can lend elegance and variety to commercial spaces and homes (Figure 8-2).

Courtesy of NDI–Natural Decorations, Inc

FIGURE 8-3: Floral designs are the best accessories within a home because they bring nature indoors.

A permanent botanical floral design, plant, or plant combination is as much a decorative accessory as a sculpture, lamp, or fountain (Figure 8-3). Some argue that permanent plants take the place of their live counterparts. If they did, people would probably use a live plant in the first place. There are reasons why they do not. Decorators choose to use permanent botanicals because they like the way natural plants and flowers appear and would rather use forms mimicking nature than other objects for decoration such as a stack of books or a table clock.

In the realm of horticulture, permanent botanicals are a natural offshoot of what the interiorscaper must accomplish; adding beauty to a space with the use of plant materials. Interior plant specialists are the best candidates for the design, installation, and maintenance of permanent plant and floral displays.

Who better than a horticulturist to breathe life into permanent materials? A person who has worked with plants and flowers for many years understands how a flower blossoms, how a stem bends away from gravity, and how a leaf unfurls itself, bending toward the light. Many of the best permanent botanical floral designers have worked with live materials first as gardeners,

FIGURE 8-4: A permanent plant may be a good choice when the live counterpart cannot handle available conditions indoors.

interiorscapers, and floral designers, learning how to make faux botanicals appear more realistic. They work in the product-development phase, when individual flowers, stems, leaves, and plants are designed. Some horticultural designers use these products to create floral designs, large-scale planters, and entire spaces such as themed restaurants and casinos interiorscaped with faux botanicals.

Permanent flowers and plants require reshaping so that they lose the "just out of the box" look, conforming to their shipping package, and gain a more resplendent and lifelike appearance. Taking permanent wired leaves, stems, and other floral parts within the hands and gently bending is referred to as fluffing, breathing life into artificial plant materials.

Why Use Silks?

There are many reasons why an interiorscape firm may sell and install permanent botanicals. One reason is that a design specification may call for a look lush with tropical foliage, but available light levels will not support the toughest, low-light, live plants for more than a few months. Some clients ask about faux botanicals when plant maintenance is not available or not desired. Most of the time, silks are selected because they act as a trompe l'oeil, a trick to the eye (Figure 8-4).

Interiorscapers can enjoy working with faux botanicals. They offer many of the same creative and service opportunities as live plants do. Of course, these plants do not need water, but they still require maintenance activity and replacement in order to keep them looking good. Silk plants can get dusty over time and can accumulate greasy residue when placed near food production areas. Just like with live plants, if a trash container is not placed nearby, people will use plant containers as a trash receptacle. Passers-by can be prone to grab the leaves and stems of artificial plants to see if they are live and thus misshape or damage them. Just because they are artificial does not mean they have an indefinite display life. Strong sunlight can bleach out pigments over time, requiring replacement.

Because they are artificial, clients tend to forget they are in place and overlook them. Drunken revelers can fall on them at special events or when they are placed near hotel bars and restaurants. Simply re-standing them or propping them against the wall does not do the trick. Because of these and other issues, interiorscapers should write maintenance contracts for permanent plant material installations, the chief differences between live plants and silks being less frequent visits and less frequent replacement.

Sales and design of silk plants generate revenue you might not otherwise have received. Many clients are attracted to silk floral displays as part of a live plant installation. Colors associated with a modern interior palette are not readily available in live plants. Many interiorscapers specify permanent trees for malls, major shopping centers, and other large-scale interiors and underplant them with medium and smaller size live plants. Although the initial investment is great, guaranteed plant replacement for the larger, more expensive trees is a non-issue for plants that cannot die. Permanent plants can replicate plants impossible to maintain indoors such as oak or boxwood (Figure 8-5).

Courtesy of NDI–Natural Decorations, Inc

FIGURE 8-5: Seasonal plants are more easily kept on display when designers specify silk trees. This permanent botanical boxwood sports juvenile and mature growth.

When a designer desires to create themes, permanent botanicals are appropriate for spaces such as:

- Restaurants
    - Door decorations
    - Vestibules
    - Bars
    - Dining rooms
    - Buffets
    - Restrooms
- Offices
    - Reception desks
    - Waiting areas
    - Meeting rooms
    - Executive offices

Any place where plants or flowers would be used, a design of silk materials would also work beautifully (Figure 8-6).

Courtesy of NDI–Natural Decorations, Inc

FIGURE 8-6: A campana urn of mixed foliage plants is reminiscent of a Mediterranean terrace.

How They Are Made

The manufacturing process of silk floral materials is detailed and fascinating. Each flower and plant is handmade. Polyester sheer and sateen fabrics are used for leaves and flower petals, chosen because they accept dyes in ways allowing colors to bleed and blend, thus mimicking natural plant pigments in living flowers and foliage. Once flowers are assembled, hand brushing of additional colors creates contrast and pattern. Even talcum powder is used on some materials to provide a natural "bloom" as seen on fresh grapes and plums.

The creation of artificial plant materials starts with a live version of the plant or flower stem to be reproduced. It is carefully dissected and studied in order to clearly see its components and how they may be copied in manufactured materials.

Fabric is cut by hand to replicate each petal and leaf, and then each is given dimension with the use of a heated, wand-type tool. The head of the tool is pressed into the fabric petals and leaves to give them puckers and ripples, replicating the natural forms within the live version.

Following this process, a plaster mold is made of each petal and leaf. The plaster molds are used to create durable brass or bronze molds into which additional carving detail is made. These molds are used for the stamping/cutting process allowing for multiple layers of fabric to be cut at a time.

Molded vinyl forms the plastic foundation for many types of artificial flowers and greenery. Individual petals and leaves are then assembled into flowers and plants. Some plant materials are provided with additional hand-brushed dyes or curled with heated tools to achieve more realism.

Artificial foliage branches are sometimes mounted into dried tree trunks. This provides a combination of preserved and faux plant materials lending a natural air to an indoor plant setting. First, a dried branch or sapling is mounted within a base of foam or plaster of Paris. Once set, individual holes are drilled into the trunk and permanent botanical stems are glued into place, arranged for balance and pleasing form. Larger-scale faux trees are made using artificial trunks. The trunks of the trees are made from special materials that are molded to appear like natural bark. These trees become integral parts of the interior design, not only because they visually soften the interior, but also because they can be fabricated to house speakers, security cameras, supplemental lighting, or video monitors.

Permanent Botanical Design Techniques

Many design techniques and products are used by professional floral designers to create stunning arrangements. Acrylic water, made from resin, added to

FIGURE 8-7: Realism is the key with today's permanent florals. Note the stigmas, stamens, buds, and water, all of which add to the illusion many designers demand.

FIGURE 8-8: Tulips, lily grass, and *Hydrangea* are grouped together within a design, reflecting the zoning technique.

containers after placement of stems, creates the look and appeal of water. This product is best used in glass vases. The manufacturing process of this type of design requires excellent floral design skill along with a very clean environment. If debris falls into the acrylic water prior to curing, it will become a permanent part of the design and the illusion may be ruined (Figure 8-7).

Some floral designs show the result of having focus taken away from silk flowers thus softening the overall look, giving viewers a more complex horticultural composition. Techniques such as veiling where fine-textured foliage or filler-flowers are placed over the top of larger mass flowers such as roses or sunflowers, like a veil covering the face of a bride, create a layering effect that appears more natural. The use of lines of confusion with artificial plant materials resembling twigs or roots creates a wild, just-harvested appearance (Figure 8-8).

Preserved Plants

An exciting way of presenting both small- and large-scale plants is through the use of preserved plants. Through a chemical process, plant parts are preserved by replacing cellular moisture with ethylene

FIGURE 8-9: Preserved and flame-retardant-treated plant materials have years of display duration only requiring periodic dusting.

glycol and dyes. The replacement preservation process takes some time but produces live-preserved foliage, which maintains a supple quality. Plant parts can be re-assembled resulting in entire trees with the look of living plants. Due to their flammability, preserved foliage and other preserved plant parts used in commercial environments must be flame-retardant (Figure 8-9).

FIGURE 8-10: Interiorscapers can provide beautiful holiday decorations for commercial and residential spaces, especially for existing plant-care clients.

Christmas and Seasonal Décor

A natural progression for interiorscape companies is the provision of holiday decorations. Most offices want Christmas decorations, especially Christmas trees on display in prominent areas with high foot traffic. Because interiorscapers already work with "trees," they are a likely choice for installing Christmas trees and seasonal displays. Some interiorscape companies that offer Christmascape and other holiday decorating products and services remark that such work constitutes up to one quarter of their annual income (Figure 8-10).

FIGURE 8-11: Display staff can design seasonal settings that can be fabricated, installed, dismantled, and stored. Each successive year, components can be rearranged or refreshed for a different look.

Christmascaping and other holiday decorating for Easter and special work for seasonal displays can become a separate department for an IP company (Figure 8-11). Its needs are different due to more intensive design work and short duration of installation, display, and removal. IP companies that sell Christmascapes should promote seasonal displays throughout the year in order to increase revenue and strengthen the decorating portion of the business.

Products

There are many products available on the wholesale market, and there are also those that can be fabricated in-house as part of the seasonal décor palette. Artificial floral wholesalers can supply interiorscapers with Christmas trees from inches in height up to 30 feet tall (Figure 8-12). Other seasonal décor such as wreaths, garlands, and swags are just some of the basic possibilities to which innumerable decorations can be added. Nearly all of these faux evergreen items are made from polyvinyl chloride (PVC), a type of plastic that can be rolled into sheets. Green PVC strips are creased onto a wire, the edges fringed, and then these are spun on a rotor and become brush-like, resembling an evergreen branch. Miniature lights are also added onto branches at the manufacturer level eliminating labor for interiorscape crews.

FIGURE 8-12: Office buildings, hotels, and large-scale commercial spaces have the need and budget for massive tree displays.

Services

Concept design for interior seasonal displays takes time and creativity, but good planning is worth the effort. It is best to create a long-term business relationship with clients because effective holiday displays cannot be thrown together in a few weeks. Start working on Christmas decoration planning in January if not in December of the previous year! While existing decorations are in place, clients and designers can get a better idea of how they would like their spaces to appear the following year.

It is natural for clients to think about seasonal displays just before the holiday, but this does not provide the necessary time for design conceptualization and sourcing to occur. Capitalize on their desires by meeting with them about what can be done this year as an initial approach or preview, but also concentrate on what will be installed the following year after planning, sourcing, and installation.

What goes up must come down. At the end of a holiday season, clients understandably become anxious for decorations to be removed and stored. The removal or "strike" is just as important as the installation, but somewhat less glamorous. On the other hand, the strike is usually quick. Questions to be answered when first working with the client include the following: When will the decorations need to be removed? Who will remove them? Where will they be stored? What storage containers/covers will be used to keep them free from dust, insects, and rodents?

Interiorscapers can strike and store holiday décor, keeping it clean and safe in storage facilities. The cost of strike labor, boxing and wrapping, and storage must be factored into the overall price estimate. When planning holiday decorations, interiorscapers should keep upcoming holidays and seasons in mind. For example, clients restless for the removal of Christmas holiday decorations would appreciate a touch of springtime during the dreary months of January and February. Creative design and suggestive selling techniques are a winning combination in the combat against spring fever.

Fire Safety

Of great importance for commercial interiors, interiorscapers and interior designers must only use materials that are resistant to fire. Designers are familiar with window and upholstery fabrics that are fire resistant. Plantscapers delving into work with permanent materials should seek out fire-retardant materials.

Flammability is not a problem when working with live plant material. Live plants do not ignite due to their water content. Some top-dressings can be problematic unless they are treated with a fire retardant. The manufacturer most often does this, but interiorscapers should be aware and seek out pretreated materials.

Fire-retardants are classified as either inherent or applied. Inherent fire retardants are mixed into the product's raw materials during the manufacturing process by the producer. Applied fire retardant is a mixed solution that is sprayed onto surfaces by the retailer, in our case, the interiorscape design staff. The process is rather messy and requires space where the solution can coat all surfaces of artificial plant materials and fully dry. The liquid is somewhat costly to purchase, but its value is obvious. The expense must be passed along to the client.

If there is ever a question about fire safety, interiorscapers should seek consultation with fire marshals assigned to the space to be interiorscaped, especially if artificial materials are used. They should be consulted prior to estimating costs. Note that this should be accomplished *prior* to sourcing products because inherently flame-retardant products cost more than those that are not treated. It is worthwhile to use flame-retardant products. Fire marshals along with building owners and interior designers most often require them. Fire marshals have the authority to close off an area because materials used in the interior design do not follow fire-safety standards. Imagine the problems associated with using unsafe materials in a permanent botanical interiorscape and having to remove everything and replace it with materials that meet fire codes.

Fire prevention professionals are interested in how products used for interior displays react to the Field Flame Test. An object such as a silk plant leaf is exposed to continuous ignition for 12 seconds. This could be accomplished by holding the flame of a lighter to a leaf. Once the flame is removed, the product must self extinguish within 2 seconds. In addition to this, any flaming particles or liquids must not continue to burn after hitting the floor.

FIGURE 8-13: Simple and colorful, a brief menu of traditional floral designs with a twist could be added to the product and service line-up of the interiorscaper.

Floral Design Services

Over the decades, interiorscapers have answered the call for the provision of floral designs as part of their product offerings. High-end clientele appreciate the finesse of a fresh floral arrangement displayed in reception areas. It is possible for some clients to purchase vases of flowers for their employees on special occasions such as birthdays, holidays, and promotions. Interiorscape technicians can create simple, contemporary floral designs and deliver them when servicing the account (Figure 8-13).

Fresh flower wholesale sources are numerous and can be found through brick-and-mortar wholesale florists or wholesale floral websites. Large cities have the benefit of entire cut flower districts where cut flower needs can be easily met by stopping off first thing in the morning to make purchases on the way to the store (Figure 8-14).

REMINDERS FOR FLORAL SERVICE

Clean containers—wash after every use with disinfecting solution.

Remove all foliage that would fall below the water line prior to arranging.

Keep arrangement out of direct sunlight for greater longevity.

Refresh design as needed; remove spent flowers and foliage.

FIGURE 8-14: A typical display of fresh flowers, in this case tulips, sold by the bunch in a large city wholesale flower market.

It is best to keep floral arrangements uncomplicated. The best thing is that simple floral designs are in style, using the elements of one type of plant material, a plain glass vase, and water. Tropical flowers such as bird of paradise and *Anthurium* are very long lasting. Even a few tropical leaves placed in a glass vase look contemporary and fresh. These designs using one type of flower are called monobotanical designs. This type of design is a great add-on sale for interiorscapers because, at the grower level, most flowers are bunched and sold in a singular variety. This contemporary look is fast to produce and long lasting if proper post-harvest practices are employed (Figure 8-15).

Service staff should remember that nothing is more unattractive than dying flowers and dirty vase water. It reflects poorly upon the client and the horticultural service provider.

FIGURE 8-15: Contemporary glassware holds monobotanical designs. The glass and the water must be kept clean and clear, so the designs should be serviced about every three days.

Postharvest Care and Handling of Floriculture Crops

When processing fresh-cut flowers and foliage, technicians should start with clean, sanitized, debris-free buckets. Commercial sanitizing solutions work well to remove bacteria and sediment from buckets and vases. At the time of purchase, select flowers that are the freshest at

FIGURE 8-16: A lily at this stage is perfect for corporate floral design. Employees and client guests will enjoy watching the flowers open over days at a time. Floral designs created with such flowers have the potential for several days of display.

market, with green, healthy foliage and no petal or leaf drop. Yellowed foliage may be a bad sign where the cut flower crop has used its carbohydrate stores. This disorder is sometimes seen in chrysanthemums, alstroemeria, and protea. Just because cut plant materials exhibit yellow leaves, it is not always a bad sign. Pale greens and yellows may be the natural color for a particular crop. Since long display life is more important than immediate, short-term impact with floral designs for commercial environments, select plant materials in tight bud (Figure 8-16).

Before flowers are prepared for hydration, mix floral preservatives with warm water (100 degrees F). The carbohydrate in fresh flower foods goes into solution faster in warm water, like sugar in warm tea. Allow the flower food solution to cool to room temperature. Room temperature water encourages slower, even hydration throughout the stems, foliage, and petals.

Processing should be accomplished quickly, upon receipt of perishable cut flowers. Re-cut all flower stems with a sharp knife. Research shows that flowers cut in the open air will last as long, if not longer, than those cut with dirty underwater cutters. It is important to re-cut flower stems to remove scabbed tissue, which is the way a cut flower naturally "heals" itself. The waxy substance produced by the stem in order to halt moisture loss is actually impervious to water.

Cut materials on a slant with a sharp knife, removing about one inch of the stem. A slanted cut allows for good water uptake because the stem does not sit squarely on the bottom of the bucket. If using floral foam for design work, a slanted cut lodges more securely into fresh floral foams as opposed to blunt cuts that can displace during delivery.

Do not let foliage fall below the water line. All surfaces of cut plant material are covered in unseen microbial growth, especially the wide surface areas of leaves. Allowing them under the water line encourages bacterial growth, which clogs water-conducting xylem tissue. It also makes the water appear cloudy or green and can smell horrible. For good floral longevity, no foliage should appear underwater unless it is for special effect in a short-term display such as for a party or reception (Figure 8-17).

Moisture and carbohydrates from phloem tissue flows from the cut end of the stem every time a stem is re-cut. Of course, it is necessary to do this to remove scabbed tissue allowing fresh flower food solution to rise into the stem. This **exudate** is potential food for bacteria. Bacteria and bacteria-produced slime in the bucket and vase water plug stem vessels. This underscores the need for fresh flower food solutions to increase cut flower longevity.

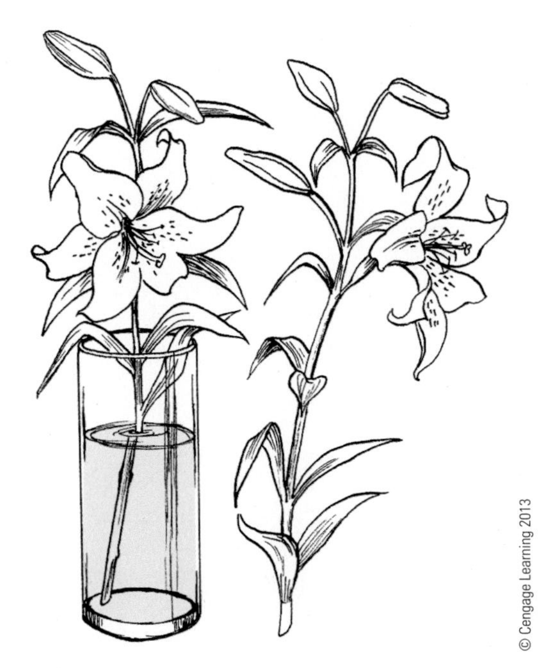

FIGURE 8-17: Prior to placing flower stems in water, whether for hydration or design, remove foliage that would fall below the water line to avoid bacterial growth in vase solution.

Fresh Flower Food

Consumers, retailers, wholesalers, and growers should be encouraged to use fresh flower foods. There are three main elements used in them, all of which are helpful to extend the display time of cut flowers.

The first main element is a carbohydrate, either sucrose (table sugar, which mixes well in water) or glucose (dextrose, which does not clump in a dry state, therefore making it easier for mixing in water), which is the "food" of fresh flower foods. Carbohydrate is the substrate used for energy to carry on life processes such as growth and floral opening. Cut flowers that need to open and bloom benefit from the carbohydrate additive; therefore, they benefit from fresh flower food. A snag is that these carbohydrates also help fungi and bacteria to grow. In order to counteract that, a biocide component, the second main element, aids in keeping bacterial growth along with fungi, yeasts, and molds from plugging-up water-conducting vessels. The third element, a biocide, usually 8-Hydroxy quinoline citrate (or 8-HQC), loses effectiveness after about 1 to 2 days.

Citric acid lowers water pH, which, ideally, should be 3.0 to 4.5, making the solution more acidic. This aids in water flow into the stem because moisture in plant material is acidic. A secondary aspect is that bacteria do not grow as quickly in acidified water.

Always follow the manufacturer's directions when mixing flower food solutions. It wastes money and does not help flowers last longer if small amounts of flower food concentrates are used. If staff flower processors use too little concentrate, the biocide and citric acid concentrations are not high enough to keep bacteria in check. Couple this with the added carbohydrate,

GREEN TIP

When a plant is removed from the interior because it has lost its aesthetic appeal, it can be maintained outdoors in a shady area, weather permitting. The plant will respond to optimal light levels and grow lush foliage again. The plant may gain back its good looks to a point that it can be reintroduced to the indoors or cut and used in floral design.

FIGURE 8-18: A rack of selected wholesale flowers ready for wrap and transport to the store. They will be made into designs and taken to accounts the same day.

and a bacterial farm is close at hand. Perhaps this misuse is one of the reasons why some people feel flower foods do not work.

Refrigeration

Refrigeration for cut flowers is a nice thing but not always necessary. A cut flower cooler removes heat energy from speeding up the overall aging process in cut flowers. The purchase cost and operation of a cut flower cooler may not be necessary for small volume floral departments if only a few accounts enjoy cut floral designs. Interiorscape management can organize purchases so that once flowers are procured from wholesale sources, they make their way to accounts within 24 hours (Figure 8-18).

Cut flower longevity is optimized at a temperature range of 33 to 38 degrees F (1 to 3 degrees C), but this range is not appropriate for all types

© Cengage Learning 2013

FIGURE 8-19: Modern cut flower coolers store as well as display stock. Displays are important if a store encourages walk-in traffic.

THE THINGS THAT SPEED UP SENESCENCE IN CUT FLOWERS

Warm temperatures
Water stress
Mechanical damage

of flowers. While roses, larkspur, lilies, and temperate-environment flowers last longer when stored at these temperatures, cut tropical and arid-region flowers like bird of paradise, anthurium, protea, and ginger last longer when stored at around 50 degrees F (10 degrees C).

Humidity

Cut flower coolers must have high relative humidity levels to keep flowers from dessicating (drying out). A relative humidity of about 90% is optimal in a cut flower cooler (Figure 8-19). Remember, the humidity in a plant cell is near 100% and that water will move from areas of high water concentration to areas of low water concentration, so water will slowly **transpirate** from a cut flower into the air within a cooler. Refrigeration humidity is optimized by the surface area of cooling coils present in the refrigerator's condensing unit (Figure 8-20). This is the major difference between refrigeration for flowers and refrigeration for soda pop. A used beverage cooler may seem like a good buy, but its lower cooling coil surface area makes its interior too dry for flowers.

FIGURE 8-20: Increased surface area of cooling coils in refrigeration brings about higher humidity levels but also increases the initial cost and operation of floral refrigeration. This is the chief difference between floral refrigeration and refrigeration for non-perishable products such as soft drinks.

© Cengage Learning 2013

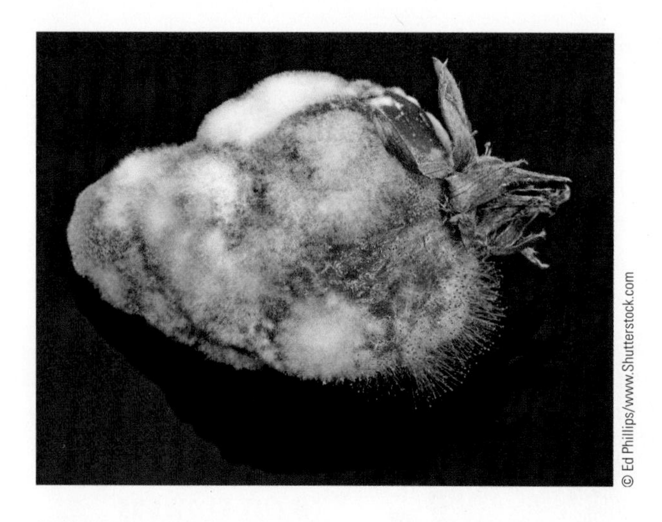

FIGURE 8-21: Gray Mold affects more than just flowers. Other horticultural products such as plants and fruits can succumb, especially in the **post-harvest** phase.

Botrytis cineria **Gray Mold**

Appearing like little tufts of white or gray cotton puffs, *Botrytis*, called Gray Mold, grows freely when moisture is present on plant material surfaces (Figure 8-21). When actively growing, it breaks up tissue integrity. Flowers may simply fall apart and leaves may actually fall off the stem because of this infection. If humidity levels are too high in cut flower coolers causing surface moisture accumulation on plant materials, *Botryis* growth will occur. You can tell when humidity levels are too high because water droplets form on cooler walls, floors, and ceilings. It is as if it is raining inside the cooler and it is better if the relative humidity is just a tad drier than that, encouraging transpiration through cut flowers into the air.

Gray Mold can be a problem with flowers enclosed in plastic packaging for long periods of time. Roses and other cut flowers benefit from plastic packaging for the short term such as in shipping from grower to retailer. Plastic film can trap moisture, providing mold with its perfect environment. It is best to provide air space around flowers, so always remove them from the packaging.

FIGURE 8-22: One bad apple or one dying flower in a bucket or floral arrangement gives off ethylene gas, which causes other flowers near it to senesce. Remove spent flowers from stock and from floral designs on display as soon as they wither.

Ethylene

Another culprit that shortens the life of cut flowers is the naturally occurring plant hormone ethylene. It is an odorless, colorless gas and is called the "aging hormone" for plant materials. When working with cut flowers, it is important to minimize the amount of ethylene gas present. The first thing is to avoid its sources found in aging flowers and foliages, particularly evergreens. Have you ever heard the expression "one bad apple spoils the whole barrel" (Figure 8-22)? This phrase is true because an aging or damaged fruit gives off exponential levels of ethylene gas that quickly age and deteriorate surrounding fruits. This also applies to cut flowers, so it is important to remove aging, damaged, or otherwise deteriorating cut flowers from the cooler, the workplace, and every floral arrangement.

The effects of ethylene manifest in one of two ways: flowers wilt, wither, and die or flowers fall off the stem. Wilting-type, ethylene-sensitive flowers include carnations, sweet peas, and many others. Some shattering-type flowers that stay intact but fall off the stem are snapdragons and dendrobium orchids.

Besides being produced by plant material itself, ethylene is also produced from exhaust fumes; smoke, including cigarette smoke; fruits; vegetables; and plant material debris. Employees must never store lunches and leftovers in a cut flower cooler.

Floral department employees must commit to minimizing ethylene production by sanitizing flower buckets, coolers, worktables, and tools. They can inhibit its action by purchasing flowers that have been treated with anti-ethylene agents at the grower level. One such ethylene inhibitor is 1-methylcyclopropene. Appropriate refrigeration as described previously is also an ethylene inhibitor.

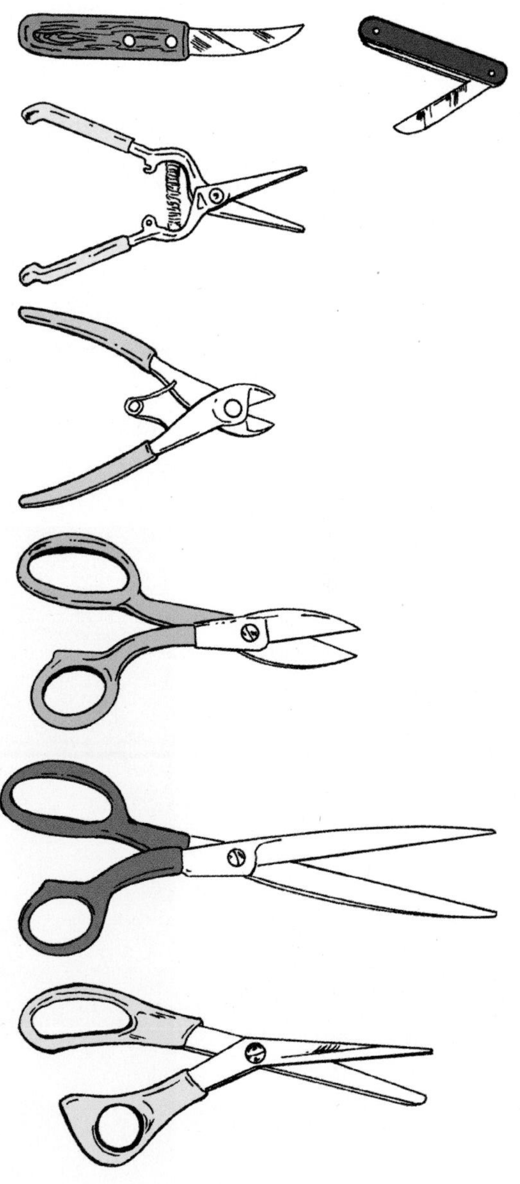

FIGURE 8-23: Floral designing tools: Cut flower knives, stationary and folding; snips for cutting small, woody stems; wire cutters, florist clippers for cutting light gage wire and small, woody stems; ribbon shears reserved for cutting ribbon only and utility scissors.

Tools and Mechanics

Floral designing requires a different set of tools than interiorscaping. They are special for the work of cutting flowers and other materials associated with floral design (Figure 8-23). All too often, the right tool is used for the wrong job. Never use a cut flower knife to cut apart a cardboard box. It dulls the blade quickly. Designers should reserve quality, sharp scissors for the sole purpose of cutting ribbon, important for Christmascape and other design work. Aggravation sets in when the only available pair of scissors does nothing but chop and fray ribbon ends. Poor tool usage cuts down on productivity.

Floral designs can get more complicated than just adding fresh flowers to a vase of flower food solution. This requires the use of floral design **mechanics**, the items used to secure placements within a floral design. Mechanics may be decorative or non-decorative.

Sound, solid mechanics are essential for physical balance of floral designs and require practice to master. One of the most popular forms of floral design mechanics is fresh flower foam. It is a resin-based product with a cellular structure similar to that of the interior of a fresh flower stem. It is very soft and absorbs water readily. When hydrating fresh flower foam to create a floral design, simply drop the foam

FIGURE 8-24: Do not force fresh floral foam below the water line. Allow it to free-float until saturated, which takes about 30 seconds for a standard-sized brick. In order to allow for horizontal floral stem insertions, allow the brick to be about 1 to 2 inches above the rim of the container.

on the surface of a basin of fresh flower food solution. Allow the foam to soak up the solution at its own rate of speed, usually about half a minute, until it is fully saturated. Never force the foam shape under the water line to hasten hydration (Figure 8-24). This can create what is called "dry core," areas within the foam's interior that do not absorb moisture. If a stem is inserted into a dry core area, it will wilt.

Of use for interiorscapers with clientele preferring clear glass, taped grids and kubari of various media provide helpful solutions to creating effective floral design mechanics. Sometimes, it seems impossible to keep flowers in place in a glass container. The mouth of the container may be too wide, necessitating the use of more plant material than what is available.

A grid work of florist anchor tape or transparent adhesive tape, which is more easily hidden, can be made over the mouth of the vase (Figure 8-25).

FIGURE 8-25: Grids made of florists' waterproof tape or transparent tape aid in holding stems in place, thus necessitating less floral material to achieve a desired form.

Take care to remove residues and dry the vase thoroughly before application. Once the grid is in place, add fresh flower food solution to the container to within about 1 inch of the rim. Once flowers are in place, add additional foliage to conceal the grid.

A second type of mechanic is called *hana kubari*, Japanese for "flower holder" (Figure 8-26). Kubari can be made with numerous types of materials. Some popular versions involve a network of fresh willow with all foliage removed. The flexible branches of willow or similar plant material can be wadded up and placed within a container. Floral wholesale supply companies sell heavy gauge decorative wire in many colors. It can also be formed into shapes conforming to the interior of a vase. It is reusable.

FIGURE 8-26: Kubari have been used throughout antiquity as devices to fix flowers in place. With some practice, these mechanics can be as creative as the completed design.

SUMMARY

The business of interiorscaping can open doors to other areas of creative expression. When delving into designing with permanent botanicals, holiday displays, and fresh floral design, sales staff, designers, and installation technicians should keep in mind the theme and effect of the finished project. It is always best to listen to the stakeholders in order to deliver exceptional products and services.

Observation of other settings can provide valuable information in learning about complementary design products and services. It is valuable to view the work of other companies in different cities and geographic regions to get new ideas. Ask yourself what makes it great or what makes it unsuccessful.

Ultimately, with hard work and time, interiorscape designers develop their own sense of style, which reflects upon the good name of their company. In order to be a good designer and feel comfortable with your work, you must practice.

SECTION THREE

SCIENCE

Thus far, we have looked at indoor plants as elements of design, exploring the benefits of color and textures, and how we can use them to provide elegant proportion to a room design. Unlike paintings or furniture, plants are living things with requirements of water, light, and nutrition in order to grow and reproduce. When we study plants, we study their life processes and predict how they will behave in a given time and environment. This section of the textbook will relate plant science to the culture of indoor plants.

CHAPTER 9

Origination

INTRODUCTION

This chapter embarks on the science of interior plantscaping and explores indoor plants from the standpoint of being living, respiring organisms. There are many types of plants on the planet, but they can be organized in simplistic categories based upon their make-up and reproduction. The relationship of plants and people within the indoor environment gives us cause to look more closely at their interaction and how plants provide important benefits.

FIGURE 9-1: Moss growing in a jar, an alternative plant display.

Vascular and Non-Vascular Plants

Vascular plants contain xylem, to conduct water and nutrients up the stem, and phloem, tissue that conducts water and carbohydrates down the stem. Nonvascular plants such as liverworts and mosses do not have xylem and phloem tissue. They are beautiful but challenging to keep alive as potted plants indoors because they dry out so easily. Plantscapers use various types of mosses, displayed in a dry state, most often as a surface "planting" to cover the mechanics of grow pots and soil mixes within display pots. Tender liverworts and mosses can be cultured in terrariums (Figure 9-1). Although non-vascular plants are often avoided in commercial settings where more durable genera such as *Dracaena* or *Sansevieria* provide better value, some high-end accounts would appreciate them. The critical point of tender plant display is consistent maintenance of aesthetic appeal. All plants must remain fresh and beautiful, working with the ambience of the interior space, not against it. If plant maintenance time or availability becomes an issue, resilient plants do a better job in staying handsome longer than do delicate plants.

Ferns

Vascular plants are divided into those that flower and those that do not. Ferns are nonflowering, vascular plants that reproduce via the distribution of spores. Many indoor plants are called ferns, but they are not true ferns. The simple test is that if a plant is in flower or has a fruit (the product of a fertilized flower), it cannot be a true fern.

Ferns reproduce via spores (Figure 9-2). A **sporophyte** (a spore-bearing plant) displays brownish or blackish structures on the undersides of its leaves. These structures are not spores but are **sori** (singular **sorus**), and spores are borne within them. The sorus is like a miniature shield that protects **sporangia**, minute structures shaped like covered soup ladles. When sporangia unhinge, they release lightweight spores. In short, spores, the reproductive body of ferns, are produced within sporangia, within sori. They are essentially the equivalent of a seed in flowering plants, except a spore does not germinate into a young plant. It does something different.

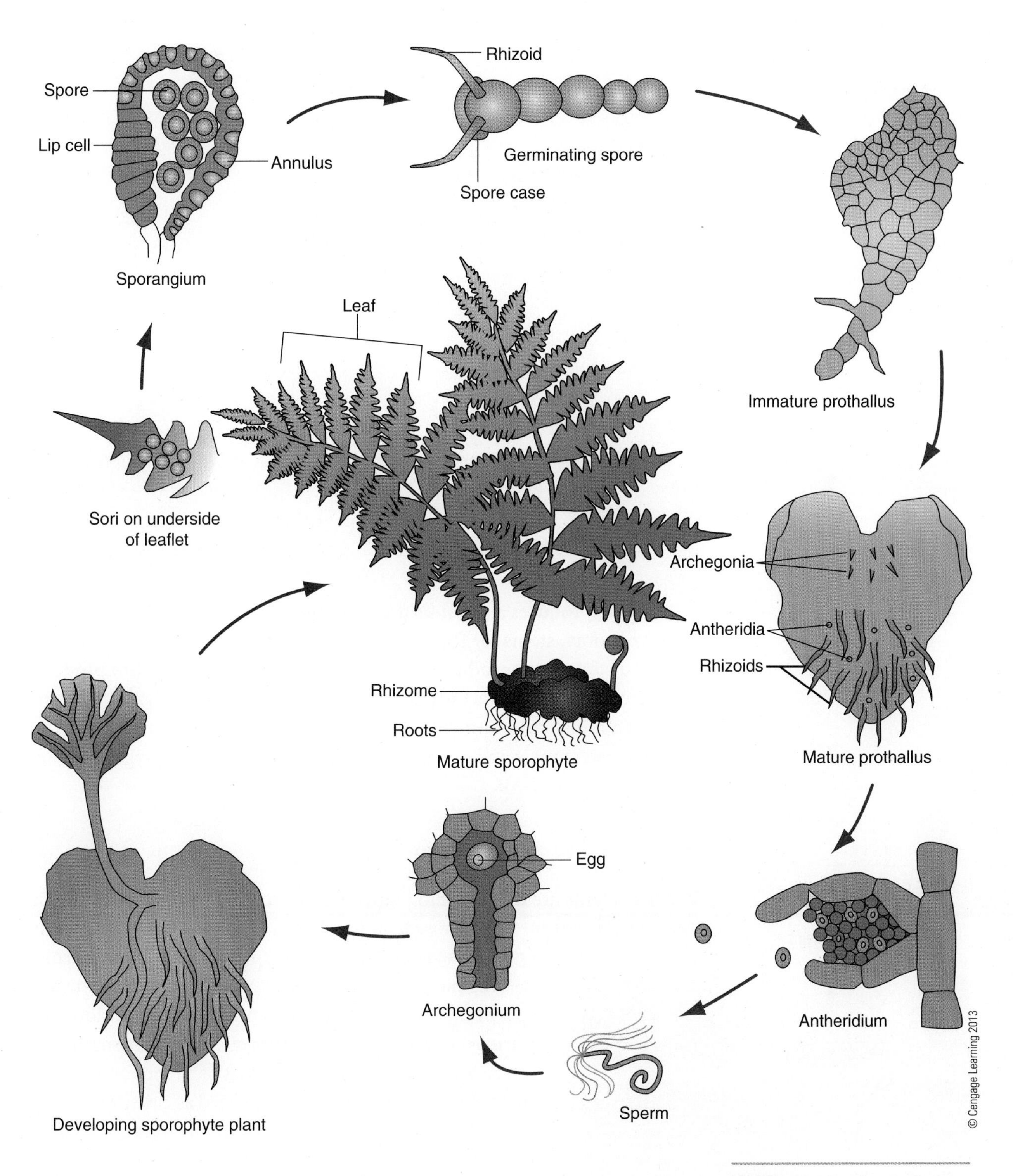

FIGURE 9-2: The life cycle of a fern.

FIGURE 9-3: As long as there is moisture and bright light, a mature sporophyte can grow and it is not uncommon to see them flourishing on a brick wall.

A spore produces a **prothallus**, a delicate, photosynthesizing plant part absent of roots, stems, or leaves. The prothallus may bear one or both sex organs; **archegonia** produce the sperm cells and the **antheridia** produce the egg cells. Some ferns only produce archegonia or antheridia, but not both. In a moist environment, sperm swims from the antheridium to the archegonium to meet and unite. The product of this union is a zygote, a baby fern plant (Figure 9-3).

Gymnosperms and Angiosperms

Vascular plants are divided into plants that flower and produce seeds in an ovary (**angiosperms**) and those that do not flower but produce naked seeds (**gymnosperms**). Pines and spruces are gymnosperms, bearing naked seeds in cones. *Cycas revoluta* is a gymnosperm that is commonly grown indoors or outdoors in warmer climates (Figure 9-4). Its native habitat in southeastern Asia and Japan is similar to that of an indoor environment. Angiosperms are further defined as either monocots or dicots, the differences shown by the number of seed leaves first appearing after germination. Monocots produce one leaf upon germination while dicots produce two. As evidenced in mature leaves, monocot leaf veins are arranged in parallel lines and dicots exhibit a network of veins.

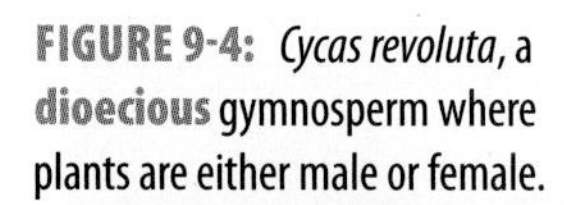

FIGURE 9-4: *Cycas revoluta*, a **dioecious** gymnosperm where plants are either male or female.

Evergreen Perennials

Most plants used by interiorscapers are evergreen perennials, meaning that the plants retain their leaves year-round and the plants have the potential to live for many years. *Dracaena*, *Ficus*, and *Dieffenbachia* are all examples of popular evergreen perennials. They have the ability to flower usually once per year, this occurring on mature plants. Many people are amazed to see interior, tropical plants, normally grown just for their foliage, begin to flower. Again, flowering of such plants tends to occur in mature specimens and where plants are somewhat root-bound within pots.

Biennials and Annuals

Biennial plants produce only foliage in their first year. In their second year, they flower, produce seed and die, thus their name which means "two years". Sometimes, plants that have a short life span (one season) are used indoors due to their unique color splash and seasonal theme. *Cineraria* (Figure 9-5) and *Calceolaria* are two popular annual plants grown in pots and used for temporary color. Note that these plants germinate, flower, produce seed, and die in one season. Many annuals grow and flower repeatedly outdoors during the spring and summer months in North America. The same plants, if kept indoors, do not have as long of a life span mostly due to lower light levels. It takes high light intensities to keep a plant in constant flower.

FIGURE 9-5: A mass of springtime color from a mass garden display of Cineraria.

The best way to understand the culture of a plant is to know from where it originated. Most plants that are used for interior displays are from within or near the Tropics, defined as the region between 23.5 degrees north and 23.5 degrees south of the Equator (Figure 9-6). The northern boundary is the Tropic of Cancer, which lies at latitude 23° 26' 22" north of the Equator while the Tropic of Capricorn lies 23° 26' 22" south of the Equator. These are rather rigid measurements, but there are numerous climate differences within the Tropics due to geography and global climate changes. Plants growing at higher elevations are used to cool temperatures, while plants in lowland areas may require more water. The effects of global climate change are just coming to light as studies are conducted to find the effects of a widening tropical zone.

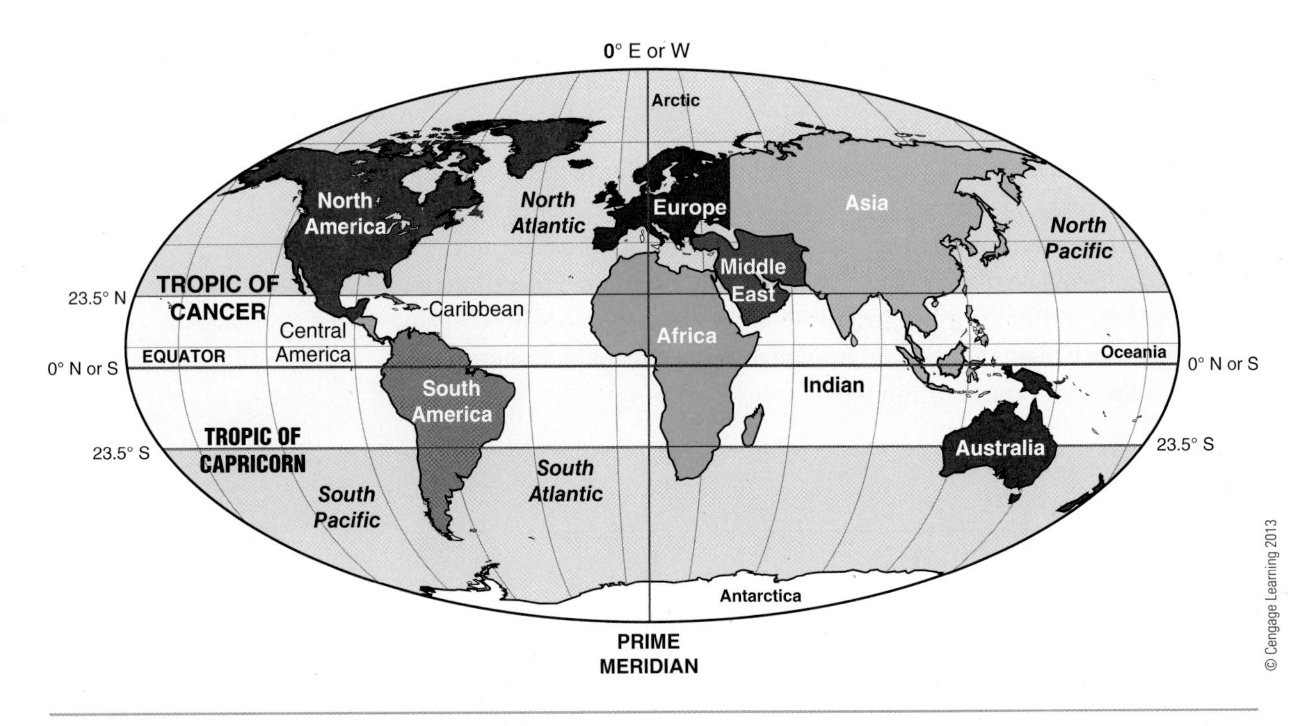

FIGURE 9-6: The Tropics, all of the earth between the Tropic of Cancer and the Tropic of Capricorn.

In general, the climates from which most houseplants arise are warm and moist with ample water in liquid and vapor forms. These regions have produced plants that not only are easily adaptable to the interior but also are attractive due to lush growth habits and interesting foliage. Most indoor plants would be "understory" if they were in their native habitat, meaning that they would normally grow in the dappled shade found under taller plants (Figure 9-7). Within these tropical forests, organic matter would decompose quickly, hastened by warm temperatures, abundant moisture, and microbial activity. This action creates soil that is rich in nutrients and high in air-filled pores. Before getting carried away with this *Jungle Book* theme, not all regions within the Tropical Belt are rainy and steamy. Some are arid, with little annual rainfall but profuse sunshine, producing plants that rely on small amounts of moisture and abundant light.

FIGURE 9-7: Plants of various genera compete for light as understory plants, whether in a botanical garden setting or in their native habitat.

How Are Interior Plants Different?

Indoor air temperatures from 68 to 75 degrees F provide a comfortable living/working environment for humans and plants. Tropical interior plants can tolerate humidity levels that are lower than those of their native habitats. They can maintain their beauty at indoor light levels. This combination of similar affinities is the chief reason why we can bring plants indoors, even though they were never designed for such a purpose! Any plant that can live indoor year-round is a houseplant. Knowledge of the geographic and climate conditions of a plant family can go a long way toward keeping its members healthy indoors (Figure 9-8).

Interiorscape plants are different from those that have lived outdoors all their lives. Plants grown for interior culture have been raised in controlled environments and provided with what they need to produce the best-looking, healthiest plant in the shortest amount of time. The limiting factors associated with indoor plant culture are light, water, and nutrition along with keeping out or at least lessening the effects of insects and disease. Overall, they have had a coddled upbringing, but they maintain their genetics tempered by natural selection of their relations in the wild.

FIGURE 9-8: Many plants thrive in environmental conditions pleasant for humans.

Poisonous Plants

Most all of the plants suited for interior culture are safe to handle and do not pose a threat to the health of people or animals. It should be considered that some people could have allergic reactions to particular plants through contact or through ingestion; for instance, sap from a plant could cause itching or a rash. It is known that all parts of *Alocasia, Anthurium, Clivia, Dieffenbachia, Epipremnum, Euphorbiaceae, Ficus,* and *Spathiphyllum* may cause severe discomfort if ingested, and contact with their sap may irritate skin. While the fruit of *Monstera deliciosa* is edible, some people may experience skin irritation if handing this plant or its fruit. These plants' cells contain raphides, extremely small but sharp calcium crystals that irritate oral and esophageal tissue. Another consideration is that many plants have yet to be screened to find out if they could be problematic. If someone comes in contact with a plant and has an adverse reaction, they should know the name of the plant or collect a sample and seek medical help.

GREEN TIP

Incorporate plants into your everyday life. Not only do they improve mood, but they also improve indoor air quality. The old saying "The shoemaker's children are often shoeless" is also true for horticulturists. Remember to keep plants in your own home.

Horticulturists should keep in mind that the multitude of plants that are offered from greenhouses and retailers are time-tested. Many of our favorite plants have been enjoyed by generations of people since the 18th and 19th centuries. They are safe for sale and display.

Plants for Clean Air

Plants are air-cleaning machines. This opening sentence is one of the hottest marketing slogans today for the green industry. If a client can afford plants, they probably already have them present in some form, whether someone in the company cares for a few potted plants or if they have an incredible installation maintained by one of the best interiorscape companies. Savvy clients know that great-looking installations improve the interior environment and pave the way for success with their clients.

What people do not know, especially potential clients, is that indoor plants do more than improve the look of a space. They actually raise the quality of the air. The planet is covered with vegetation, from tall trees to low-growing ground covers and even inconspicuous plants. When plants take in carbon dioxide for photosynthesis, they give off oxygen as a byproduct—clean, pure oxygen. Oxygen is needed by animals to live. The production of oxygen is largely a part of the leaf environment.

The earth does not hold only clean air. The air we breath consists of more than just oxygen; other elements and compounds are part present which may be detrimental to public health. What makes the situation perhaps even worse is that our time is often concentrated indoors. Professionals may spend 7 to 10 or more hours per day in an indoor work environment (Figure 9-9). A closer look at the surroundings of work and living spaces finds a set of factors that shed light on why this can make people feel sick. A phenomenon that has been researched for years but is not widely known, it has lead to a phrase known as "sick building syndrome" (SBS).

FIGURE 9-9: The workplace environment must be of the highest quality in order to help people achieve professional goals. Ramifications for productivity and well-being are at stake.

Sick Buildings

In order to conserve energy and costs, construction techniques and materials have been developed that keep heat energy inside during the winter and keep cool air within buildings during the summer. Well-sealed buildings are energy-efficient

FIGURE 9-10: Even with doors and windows closed, older constructions allowed for air exchange with the outdoors. Modern construction is more energy efficient, but air exchange may be hindered.

and do not allow for much air exchange with the outdoors. Take for instance the comparison of a house built in 2000 with one built in 1850. The windows, doors, and insulation (or lack thereof) are completely different. Nineteenth-century homes used single-pane glass windows. When the weather was warm or if inhabitants needed fresh air, they simply opened a window or door (Figure 9-10). Even when the window was closed, the lack of exacting measurements in construction processes meant small cracks and open spaces would always be present. This is not a part of the efficient products and practices in the construction of office buildings or hotels today. Many window designs have two or three layers of glass, sometimes with poor thermal-conducting gasses between that do not allow for much loss of heat to the cold outdoors. Such windows are more accurately *glass walls* that let light rays through and offer exterior views but offer no ventilation.

Tightly sealed buildings can become sick buildings. They are characterized as having a high proportion of people working or living in them who exhibit symptoms of illnesses that may diminish when they are away for a period of time. People working in sick buildings may suffer from any of the following symptoms:

- Allergies
- Asthma
- Congestion
- Headache
- Mucus membrane irritation

Some people are more chemical-sensitive than others. Certain high-risk groups include infants, asthmatic children, and those who happen to be chemically hypersensitive. Imagine the problems associated with feeling sick when you spend time in a certain place but feeling okay when you leave. If it were a work location, supervisors and co-workers may get the wrong impression, deciding that the employee was unhappy with the job position. More than likely, people who have these reactions may never know exactly why they feel the way they do, let alone the debilitating psychological effect of feeling ill while at work but feeling fine after leaving, day after day. What exactly makes people feel ill in sick buildings?

Volatile Organic Compounds

What makes the lack of air exchange a problem is that concentrations of potentially harmful airborne chemicals are constantly introduced into indoor air. These chemicals are called **volatile organic compounds** (VOCs) and are emitted from numerous products including those used to furnish interiors and those used in office products. A more comprehensive list of VOCs includes:

Adhesives
Bioeffluent
Blueprint machines
Carpeting
Caulking compounds
Ceiling tiles
Chlorinated tap water
Cleaning products
Computer screens
Corrections fluid
Cosmetics
Draperies
Duplicating machines
Electrophotographic printers
Fabrics
Facial tissues
Floor coverings
Gas stoves
Grocery bags
Microfiche developers
Nail polish remover
Paints
Paper towels

Particleboard
Permanent press clothing
Photocopiers
Plywood
Pre-printed paper forms
Stains and varnishes
Tobacco smoke
Upholstery
Wall coverings

These products may be **off-gassed** where chemical compounds are released prior to installation, but this cannot always be completed in a short time. Some products take years to off-gas. In addition to these, humans are also sources of indoor air pollution through the release of volatile organic compounds called **bioeffluents**, products of respiration. The smaller the indoor space compounded by the higher the number of people in that space, the more noticeable this concentration becomes. Bioeffluent from humans are:

Acetone
Ethyl acetate
Ethyl alcohol
Methyl alcohol

Along with

Alcohols
Aldehydes
Ammonia
Carbon monoxide
Hydrogen
Hydrogen sulphide
Indol
Mercaptans
Methane
Methyl indole
Nitrogen oxides
Phenols
Volatile fatty acids

The presence of VOCs does not adversely affect everyone. Chemical concentrations and duration of exposure can be factors. Reactions may be acute, where someone feels ill for a short time. A chronic condition could develop where a person is ill for a long period.

Indoor Plants for Indoor Air

A living plant is an entire air cleaning system. Both the leaf and root environments are important to the completion of removing harmful chemicals and microbes from the air. Research shows that the more chemicals in the air, the harder plants work to remove it, with rates of removal dropping when levels of toxins subside (Figure 9-11).

The root environment is an important region of air scrubbing, which seems odd because roots and associated media substrates are *underground*. How do VOCs enter the soil? This is accomplished through gas exchange of stomates on leaves as well as air being pulled through the soil, caused by water movement from the soil into roots. Ultimately, organic particles, whether absorbed through leaves or air movement into the soil, are used by microbes for food.

A pattern of action occurs involving the relationship of water and air. Water in the soil of a potted plant moves from the medium into roots and throughout the plant. As water moves out of the soil, it is replaced by air, pulled from above the soil and into the media.

The air moving into the soil contains oxygen, which is needed by soil microbes to live and continue their work. Part of the work they perform is to take in nitrogen, also found in the air, and convert it to nutrition used by the plant. Soil microbes break down organic particles like decayed leaves and other plant parts in the soil, which are used by the beneficial microbes and plants for food. It is a wonderful relationship.

Earlier in this section, we referenced the fact that the leaf environment is able to exchange gasses with the openings of stomatal pores. An important

FIGURE 9-11: As plants remove chemicals from the air and toxin levels subside, the plants adjust their rate of removal.

function of this action is that water vapor is also exchanged and balanced between the plant and indoor air. People and pets enjoy healthy indoor humidity levels between 35 and 65%, while percentages above this encourage mold and mildew growth. Dry indoor air places stress on sinus passages, causing the potential for infection. Low humidity levels are also detrimental to wood furnishings, which can dry and crack. High humidity levels can result in costly repairs; for example, condensation accumulating on glass can damage windowsills. Plants have the ability to balance indoor humidity levels, adding moisture to dry air as well as decreasing humidity when it is too high.

Plants suppress airborne mold spores and bacteria. Rooms with robust interiorscapes have 50 to 60% fewer airborne molds and bacteria. This built-in immunity is important for the plant because it helps it fight off infection that would be present in the air. Phytochemicals, chemicals produced by plants, that help them fight infection also help humans who in turn benefit from lowered levels of ambient mold and bacteria.

We owe a debt of gratitude to the research conducted by B. C. Wolverton at the John C. Stennis Space Center, Picauyne, Mississippi. This research sought to find ways of filtering used air and water in the creation of lunar stations. In his research, he listed the top plants for clean air based upon the ability of plants to remove VOCs and balance humidity, and their ease of maintenance indoors:

Aechmea fasciata (Silver Vase Plant)
Aglaonema commutatum (Chinese Evergreen)
Aloe vera (Aloe)
Anthurium andraeanum (Anthurium)
Araucaria heterophylla (Norfolk Island Pine)
Azalea sp. (Azalea)
Begonia x semperflorens (Wax Begonia)
Brassaia actinophylla (Schefflera)
Calathea sp. (Calathea)
Chamaedorea erumpens (Bamboo Palm)
Chlorophytum comosum (Spider Plant)
Chrysalidocarpus lutescens (Areca Palm)
Cissus rhombifolia (Grape Ivy)
Codiaeum variegatum (Croton)
Cyclamen persicum (Cyclamen)
Dendranthema grandiflora (Florists' Chrysanthemum)
Dendrobium (Orchid)
Dieffenbachia seguine (Dumb Cane)
Dracaena deremensis 'Janet Craig' (Janet Craig Dracaena)
Dracaena deremensis 'Warnecki' (Warnecki Dracaena)
Dracaena fragrans 'Massangeana' (Corn Plant)

Dracaena marginata (Dragon Palm, Red-Margined Dracaena)
Epipremnum aureum (Pothos)
Euphorbia pulcherrima (Poinsettia)
Ficus benjamina (Weeping Fig)
Ficus elastica (Rubber Plant)
Ficus maclellandii 'Alii" (Ficus Alii)
Gerbera jamesonii (Gerbera Daisy)
Hedera helix (English Ivy)
Homalomena sp. (King of Hearts)
Kalanchoe blossfeldiana (Kalanchoe)
Liriope spicata (Liriope)
Maranta leuconeura (Prayer Plant)
Musa sp. (Dwarf Banana)
Nephrolepis exaltata 'Bostoniensis' (Boston Fern)
Nephrolepis obliterata (Kimberly Queen)
Philodendron domesticum (Red Emerald Philodendron)
Philodendron scandens (Heart-Leaf Philodendron)
Philodendron selloum (Tree Philodendron)
Phoenix roebelenii (Dwarf Date Palm)
Rhapis excelsa (Lady Palm)
Sansevieria sp. (Snake Plant)
Spathiphyllum sp. (Peace Lily)
Syngonium podophyllum (Nepthytis)
Tulipa sp. (Tulip)
Zygocactus sp. (Thanksgiving Cactus)

There are still arguments as to just how many plants are needed to provide "fresh air" to a room. Some believe that one large plant such as an 8- to 10-inch potted *Dracaena* is perfect for a 100-square-foot office. Others argue that it would take three times that figure to compensate for lack of air replacement in a new office space with fresh paint and carpet. Some research states that the presence of plants indoors does not have much to do with the overall improvement of indoor air quality because experiments have worked with controlled spaces more than actual offices. Perhaps this does not replicate continual emissions as generally seen in indoor environments.

Because this segment of investigation is new, it may be premature to believe that keeping plants indoors improves indoor air quality. Interiorscapers should remember the needs of the client, people who love beauty and feel happier and more productive around plants. In one study, productivity was measured by administering a computerized test to subjects in two different groups; one in a computer lab without plants and one with plants present. The results of the study showed 12% improvement in reactions to computer tasks in the plantscaped lab. The subjects had lower blood pressure readings and felt more attentive to their work than did those in the lab without plants.

Keeping their environment balanced between aesthetics and physical and emotional health can be aided by the selection and placement of floriculture offerings. This will ensure plants have their place. After learning about all the potentially wonderful things keeping plants indoors can do to keep people healthy and happy, can there be any reason why the products and services of an interiorscaper are not necessary in today's working and living environments?

SUMMARY

The world of indoor plants is our starting point, the product of our horticulture business, the objective of plant care service, and the medium for our designs. Some of the most tenacious plants on earth, many indoor plants are evergreen perennials and are part of what we consider to be the earth's lungs. Twenty-four hours per day, powered by light energy and fueled by self-made energy, plants take in CO_2 and release O_2 that humans need to live.

Through the beneficial relationship of plants and soil microbes, organic particles in the air are absorbed by leaves, translocated through plants to the root zone, and used for energy by soil microbes. VOCs move into the root zone through water-air movement into the soil, and are thereby used by soil microbes for food. Plants' leaves release water vapor and keep indoor humidity levels more comfortable for humans and pets.

Nothing can provide the softness and beauty of a plant displayed indoors. It is a living reminder of nature, their indoor culture a true feat of science and art.

CHAPTER 10

Light

INTRODUCTION

The sun is the largest object in the solar system, and from it light and heat energy are derived. Plants and humans use solar energy in different ways, but it is essential for both kingdoms to exist. Light is most often the greatest limiting factor for the culture and maintenance of plants indoors, and interiorscapers should learn about it, from the cellular level to the entire plantscaped environment.

How Plants Use Light

One of the first things an interiorscaper accomplishes when contacted about an installation is to travel to the site to learn how much light the plants will receive. In most cases, light levels are usually low, necessitating an installation drawn from a smaller pool of plants. Occasionally, light levels are high enough to allow for leaf and shoot production. High light levels provide a different palette of plant materials such as flowering plants, succulents, and others native to sunny habitats. There are many factors involved in how plants use light. Light energy is necessary for life and its conversion into sugars or other organic molecules known as **photosynthesis**.

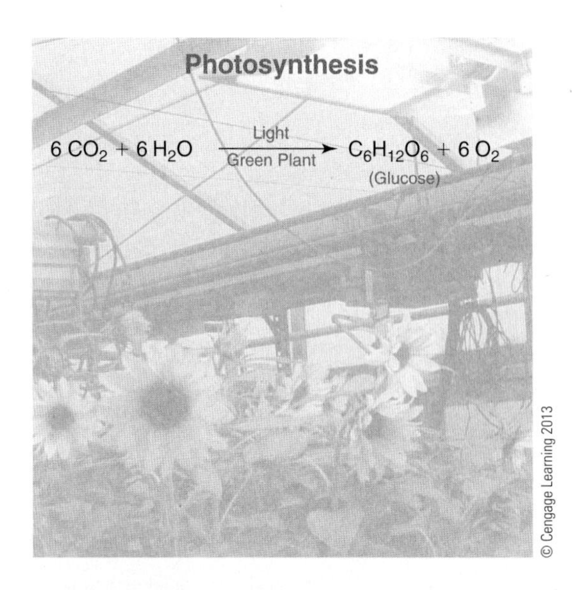

FIGURE 10-1: The chemical equation for photosynthesis.

Photosynthesis

Imagine being so busy with a project that you do not want to stop to eat. If you had the power of the Plant Kingdom in your veins, you could point your elbow out of an open window, receive light energy, and be provided with the necessary link to nourish yourself. Humans cannot do this of course, but plants can. Their leaves are powerful factories and, with just a few simple ingredients, produce carbohydrates that fuel their life processes. The production of carbohydrates is the basis for plants being able to make their own food, but they can only do so when the right ingredients are present. Carbon dioxide is a chief ingredient and is taken from the air. Water, the second chief ingredient, is taken up by roots in liquid form or by stomata, specialized pores on the undersides of leaves, in vapor form. Light must be present for this chemical reaction to occur. When there is no light, there is no chlorophyll production and, eventually, a plant will die. An important result is that plants make pure oxygen from this process which then becomes available for the animal kingdom to utilize. This process is known as **photosynthesis** (Figure 10-1).

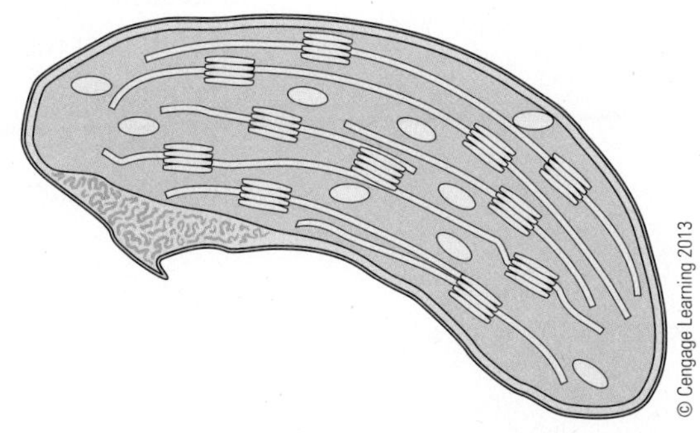

FIGURE 10-2: Chloroplasts are organelles within a cell. Dense stacks of grana hold chlorophyll molecules, which are activated by light and aid in food production.

Chloroplasts

Specialized cells named **chloroplasts** are the site of photosynthesis (Figure 10-2). Chloroplasts contain chlorophyll, a green pigment that takes in light and makes it into energy. Intense light is able to penetrate a thick leaf surface while lower light intensity cannot. The plant responds to the light intensity in its environment by producing leaves that are efficient in bright light, with thicker **palisade layers** that

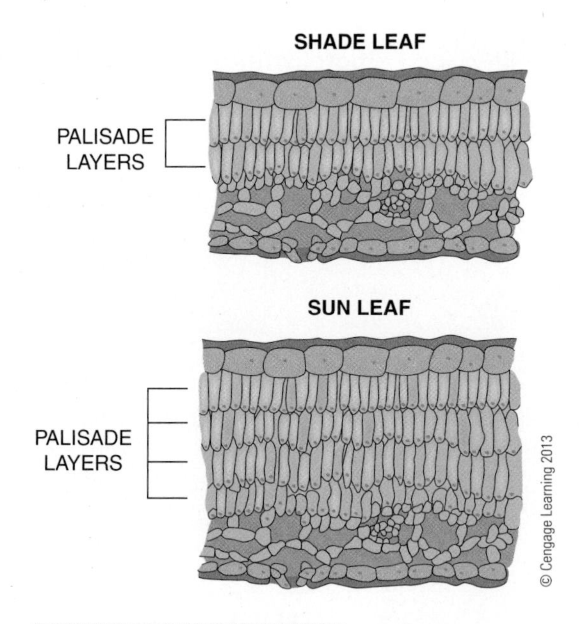

FIGURE 10-3: Cross sections of leaves demonstrate efficient use of light energy.

are well-organized photosynthesis factories (Figure 10-3). Leaves growing in bright light are characteristically thick. In lower light, the palisade layer is thinner, but broader, so that as many chloroplasts can be oriented directly to available light. Leaves produced from plants growing in the shade have broader surface areas but are thinner than their sun-grown counterparts.

Respiration

Respiration is the process in which a plant uses the food it has produced to carry on its life processes. It creates energy by using sugars produced from photosynthesis. It takes a great deal of energy to accomplish all of its work, moving sugar molecules throughout the plant, absorbing minerals, and synthesizing and using hormones, proteins, and fats.

Respiration occurs constantly, but at varying rates. For instance, respiration rates slow when a plant is in a cooler room as opposed to a warmer room. Photosynthesis, which provides the food, stops when the plant is in the dark, such as during nighttime or when lights are turned off in an office. Photosynthesis must occur at rates that exceed respiration, or there will not be enough stored carbohydrate and the plant will die in time. Table 10-1 provides a comparison of photosynthesis and respiration.

Respiration rates are greatly influenced by temperature, both the temperature roots experience in the soil environment and leaves in the open air. When the temperature rises, respiration rates are faster and the opposite is true; cooler temperatures slow respiration rates. This does not occur in a linear fashion, however. Even small temperature changes can double or triple respiration rates and this can have immense effects on the plant growth, plant performance, and overall looks. This is referred to as the Q_{10} effect, which is defined as the speed by which any chemical reaction rate increases or decreases when subjected to a 10°C temperature change.

TABLE 10-1: Photosynthesis and Respiration Comparisons

Photosynthesis	Respiration
Sun's energy used	Energy released
Food manufactured	Food consumed
Carbon dioxide needed	Carbon dioxide given off
Sugars and starches produced	Carbon dioxide and water produced
Requires light	Occurs in light and dark
Chlorophyll must be present	Occurs in all cells

Delmar Cengage Learning

Growers must produce plants in temperatures that mimic the evolutionary native habitat where the plants originated. Photosynthesis, respiration, and growth will be at optimal rates for the specific plant if the plant is grown, day and night, at temperatures similar to its native habitat. If a tropical plant is grown or maintained at low temperatures, it will not have as long a display duration as those grown at appropriate temperatures.

For practical purposes, we can transfer this information to an indoor plant site. Some businesses lower nighttime temperatures to save on winter heating expenses. A tropical plant displayed in such a space is not only subjected to the cooler nights, but light levels derived from artificial light are already low, controlled by computers, turning on at 7:00 a.m. and off at 7:00 p.m. Some systems operate by use of motion. If sensors detect movement, lighting remains on, but when workers are not present, spaces remain dark. The plant will need to compensate for the cooler temperatures by using more energy derived from carbohydrates, resulting in higher respiration levels. This is above its normal maintenance use of carbohydrates with what little was made in the low light. After a few weeks of this treatment, carbohydrate stores become depleted, leaves yellow, and root structures die as the plant tries desperately to run itself on a skimpy, shrinking carbohydrate reserve.

In another scenario, the same company may allow nighttime temperatures to rise in the summer in continued cost-cutting efforts. Cooling systems may be programmed to suspend at night and during weekends causing temperatures to rise. Again, the effect is such that the plant is continuing to respire, but at accelerated rates due to the rise in nighttime temperature. Because the plant is exposed only to limited amounts of artificial light, photosynthesis rates are the same year-round. Carbohydrate stores become depleted and the plant declines in health and appearance.

Understanding and controlling respiration rates in cut flowers is important to interiorscapers who offer floral designs as part of their retail mix. Once carbohydrates are manufactured in leaves, they are transported to storage in plant roots. When a flower is cut from the parent plant in the harvest stage, its supply of carbohydrate is no longer available. It takes energy for a blossom such as a rose or lily to unfurl its petals. Informed floral designers will add fresh flower food to bucket and vase water following the manufacturer's directions to provide a cut flower with carbohydrate in order to generate the energy to open.

Light Energy

Light is the chief limiting factor when it comes to maintaining quality in interior plantscapes. Most interiorscapers agree that if they were able to control interior light, maintaining the health and therefore aesthetic balance of plants would be easy. The problem is that most interior settings offer low light levels. The work is then to use the resources available, including the right plants for the client's budget, over time, to maintain good-looking displays.

Many people are oriented to the idea of "growing" plants in the interior rather than "maintaining" them. Growing plants indoors connotes the production of more stems and foliage—overall, a larger plant. Maintaining plants shifts the purpose away from production toward conservation of existing leaves and shoots. Generally speaking, lack of light is the greatest limiting factor of keeping plants in top form indoors. Most professionals agree it is the greatest hindrance to keeping plants looking healthy. Light levels that are comfortable for reading or other work activities are at the threshold for plants achieving growth status—the production of leaves and shoots—and maintenance status—staying as they are upon introduction to the space.

When plants grow, they need more care than when they stay the same. Growing plants need more water, more grooming, and more pruning. Plants that are in maintenance mode need some water and occasional maintenance activities such as cleaning and turning. Sites with maintenance levels of light are not necessarily bad for maintenance technicians and interiorscape companies. They may not provide gratifying levels of leaf production as well as taller plants do, but they do provide a place where healthy-appearing plants may be enjoyed and appreciated (Figure 10-4).

FIGURE 10-4: Boardrooms are important, but in some companies they are infrequently used and alternate from being brightly illuminated some days to having no light on others. Regular use of electronic visual equipment in work spaces keeps window shades and drapes closed.

Measuring Light

After a period of time, interiorscapers learn the skills of measuring light on their own without the use of light meters. They can tell just by looking at the space how various plants are going to perform. For those less seasoned to the trade, it is a wise investment for a firm to buy a light meter available for around $125. Light meters measure light in footcandles, which is the amount of light emitted by a candle in the center of a 1-foot radius sphere.

A simple and inexpensive way to read light is to place your hand about 1 foot above a white object. If you discern your fingers' shadow, there is enough light in that place to maintain a foliage plant.

Natural Light

Natural light, which is light from the sun, is the best kind of light for plants because it contains all colors. Even though sunlight may seem white or yellowish to our eyes, it contains many colors within the visible light part of the electromagnetic spectrum. Plants need various colors of light in order to maintain specific life processes (Figure 10-5).

Take into account the exposure of a space to the sun. In general, interior plants will do well when placed near an unobstructed window with an eastern exposure. This bright, indirect light is a close simulation of the dappled light plants would enjoy in tropical regions. Eastern exposures gain light from

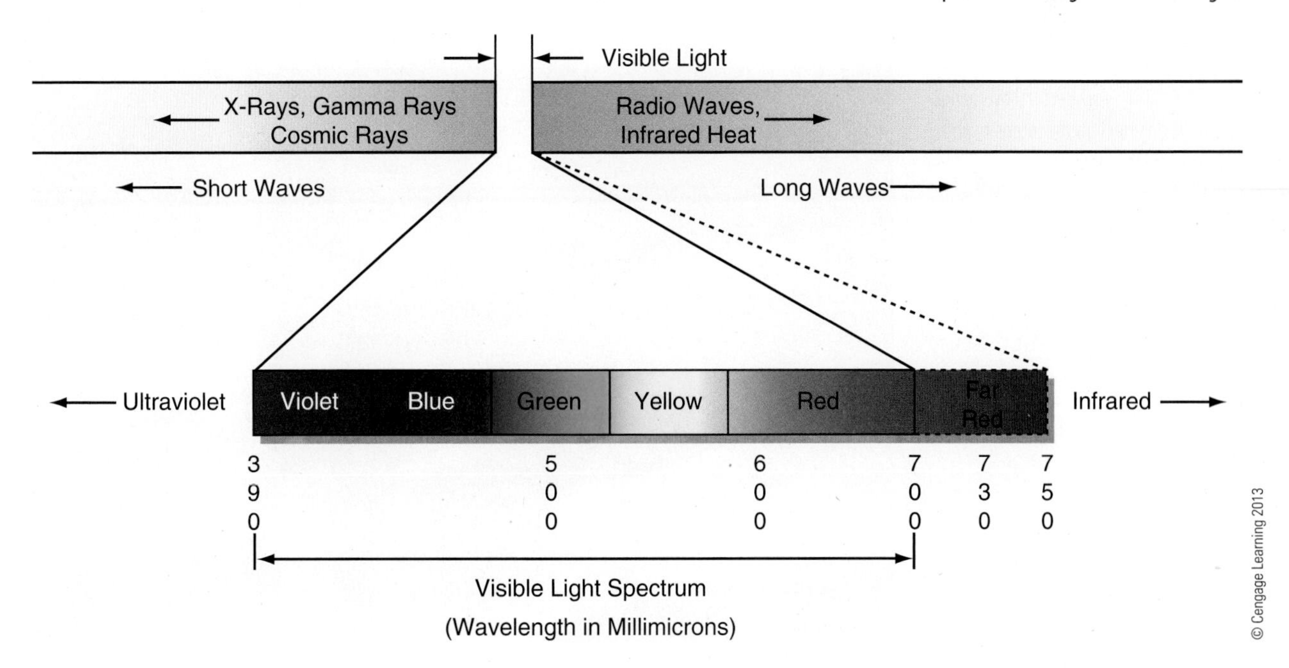

FIGURE 10-5: The electromagnetic spectrum with segment of visible light.

the morning sun when temperatures are generally cool. A western exposure would potentially have many days of bright light over time. Southern exposures provide bright light, but the sun's intensity could burn leaves. This **phytotoxicity** manifests itself in sun-scalded leaves and is an invitation for spider mites that thrive in these warm, dry conditions.

Northern exposures have, in general, the lowest light intensities of the four exposures. This does not mean that plants cannot be maintained in these spaces, however. Interiorscapers can select plants that do better in lower light as well as take advantage of plant placement closer to the light source, simply placing plants closer to a window to keep it healthy. Of course, it is not always possible to choose exclusive rooms for plant displays. A professional interiorscaper would not want to disregard a space just because it did not have the best exposure. It is possible and desirable to maintain or even grow plants in other exposures.

Intensities of sunlight are contingent upon season as well as exposure. The winter sun is lower in the sky and may therefore penetrate a room with more effective intensity. Summertime sunlight is different. The sun is higher in the sky and may not have as intensive of an effect because exterior overhangs, trees, or interior window treatments may block light (Figure 10-6).

Each of these explanations discusses light in a directional context. It is natural for plants to grow in the direction of available light. This growth is not perceptible to the human eye, but is indeed the result of plant movement called **phototropism**, or bending toward the light. This is a hormonal response within the cells of the plant. Cells on the topside of the stem elongate while those on the underside remain the same. When this occurs, a curvature

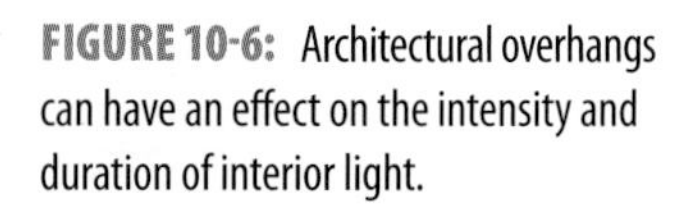

FIGURE 10-6: Architectural overhangs can have an effect on the intensity and duration of interior light.

of the stem is produced so that the newest growth is oriented toward the light source. When plants receive light from a single direction, their pots should be physically turned about 25% every week or so in order to keep the plant visually balanced.

A Word on Plant Placement

This is a good point to recall the philosophy of this text. Indoor plants provide beauty and indoor air improvement. Due to their service, they should *always* be present in spaces occupied by people to keep us healthy and happy. The duration of time that plants remain beautiful and useful should supersede considerations of whether the plant is going to live for six months or six years in the space. When a plant goes into decline and loses its visual appeal, it should be replaced with a robust version, without the client having to ask for a replacement. A major service, perhaps the greatest thing that plantscapers do, is to maintain the *aesthetic balance* of an interiorscape. This goal holds for acres of planted interior space as well as for an account with one plant placed 10 feet away from a northern-facing, tinted window in a room that just a few people use. The benefits of having plants indoors outweigh the cost.

Windows

Ultimately, natural light must pass through windows to reach interior plant leaves. The types of glass used as well as cleanliness of the windows have a great effect on the intensity of light transmitted to interior spaces (Figure 10-7). Many office buildings utilize glass that is tinted or reflective to reduce interior heat energy. High levels of tinting may result in lower power usage and utility savings, but there is a trade-off with lower light intensity for plants. Plants that do well in low and medium light levels are good choices for these environments. Conversely, plants placed close to windows with no tinting or reflectivity may be sun-scalded, especially in southern or western exposures, during the summer or in other seasons when the sun is shining for long periods of time. The damage done is actually due to the amount of heat energy rather than light intensity. Interior plants cannot stand excessively high temperatures beyond 125 to 130°F without injury.

Which do you think would have the greatest light intensity: a window with summer sun or a window with winter sun? Although the summer sun is by itself more intense than the winter sun, leaves on nearby trees may actually diffuse intense summer sun while a bright day during the winter could result in higher light intensity.

FIGURE 10-7: A glass ceiling brings full-spectrum lighting to storefronts, merchandise, and plants.

Artificial Light

Many plants living in offices or other commercial spaces rely solely on artificial light for photosynthesis. Sometimes, clients need plants in places where there are no windows, and hence no natural light. Keep in mind that artificial lights may be turned off at night, over weekends, or for longer periods during holidays. Interiorscape sales and technical staff should think about intensity and duration of artificial light when specifying and installing a live plant in such spaces. Remarkably, many types of plants can withstand such conditions and look marvelous for months, even years.

Artificial light is sometimes used as supplemental light to increase light intensity for indoor plants. This necessitates the installation of additional light fixtures by electricians. Such fixtures must be aesthetically pleasing, keeping with the overall interior design of a space, so it is optimal for the client to consult with a lighting designer in addition to the horticulture specialist. If there is not a budget for these consultants, it is best to use plants with low-light requirements. Basic light fixtures are often used by home hobbyists to grow plants indoors (Figure 10-8).

A table or entire rack using fluorescent lights can be made or purchased, keeping the bulbs just inches above the plants. This will provide the plants with intensities to trigger blooming, and because fluorescent lights are relatively cool, they will not burn foliage. The warmest part of a fluorescent light is the ballast, but its heat generally does not harm plants because it is located within the light fixture's casing.

Orchids, African Violets, cyclamen, and hundreds of other plants can be grown and continually bloom in this fashion. This type of gardening, although not very stylish in appearance, can bring a great deal of satisfaction to indoor plant lovers. When plants are at their peak of perfection, they can be placed in decorative jardinières and moved to high-traffic areas of the home for a week or two for others to enjoy. Interiorscapers should not be discouraged by the lack of aesthetics in fluorescent light racks. It might provide a challenge for product development, building a great-looking artificial light case for the culture of tender, indoor plants.

FIGURE 10-8: Most houseplants respond well to basic florescent light fixtures. They are an inexpensive way to coax plants such as these Paperwhite *Narcissus* or African Violets into bloom. Blooming plants can then be placed in common areas within the home for short periods.

The Effect of Wall Colors and Furnishings on Light

Light energy is bounced about a room, with all spectral colors absorbed except that which is reflected. The light that is reflected from a surface is the color that is seen. For instance, we see various shades of green light reflected from plant leaves because all other colors are absorbed.

Rooms that have been designed with dark colors such as browns, deep gray, or shaded reds have less light bounce than spaces decorated in light, reflective colors. Available light will be best utilized when rooms are decorated with light furnishings. White surfaces reflect the highest amount of light. Mirrored surfaces are more reflective than black, but read as gray and do have some light absorption.

FIGURE 10-9: Light-colored walls counteract the effect of dark carpet, which was selected to accommodate foot traffic. The *Aechmea* will remain colorful for months in this space.

© iStockphoto/Tiburon Studios

GREEN TIP

Use natural light to illuminate an interior. People often turn on a light switch when they could open a curtain or raise a blind. Natural light is better for live plants than artificial light.

When interior design input is possible, such as in the construction of new or remodeled space, plantscapers can suggest light colors for walls and ceilings to conserve light. It is not necessary that a room be painted white instead of chocolate, however. Interior design elements and color choices are extremely important to designers and clients. Simply put, any given plant may maintain its aesthetics longer in a lighter-colored room than in a darker-colored room, but this does not mean that the plant cannot be replaced. It will just need to be replaced a bit sooner (Figure 10-9).

Light Intensity

Light intensity is the brightness of the light made available to plants. From an applied standpoint, plants that flower as well as plants with colorful, variegated leaves need more intense light than plants with darker, green leaves. The thickness of a leaf is the result of the light intensity in which the plant was grown. Plants native to sunny habitats (*Crassula, Aloe*) often bear leaves that are thicker and are not as broad.

How brightly a light shines has an effect on the efficiency of photosynthesis. The farther away a plant is placed from a light source, the less intense is the light energy. This is compounded by the colors of the walls, flooring, and furniture of the interior. Some surfaces reflect light while others absorb it, thus lessening light intensity within a space.

When interiorscapers speak about the lack of light being a problem in using a greater variety of plants or even in keeping the old standards looking healthy, they are referring to low light intensity.

Almost all interior plants have the ability to flower, with the major exception to this being ferns. While some indoor plants are grown for the value of their flowers, from a professional interiorscaper's standpoint, they are few and far between. Indoor light levels cannot promote or sustain flowering, so blooming plants must be installed in flower and displayed for the short term. They can be thought of as a Broadway show traveling on the road. They hit town, dazzle an audience, and are gone within a short time.

Indoors, many plants do not flower (or offer just a few flowers) because indoor light intensities cannot support the reproductive phase, which requires high amounts of energy. Most interior plants are cultivated because their foliage is attractive on its own. Remember, plants that flower need greater light intensity than plants that are grown for their decorative foliage.

Keep in mind this general list to aid in plant selections in relation to light intensity.

Higher-intensity need:

- Plants that should be in flower
- Plants with variegation
- Plants indigenous to bright, sunny locations

Lower-intensity need:

- Plants that do not need to be in flower to look fine
- Plants with dark green leaves
- Plants indigenous to shady locations

Table 10-2 is a handy list of plant genera adaptable to low, medium, and high light intensities.

In some interiorscapes, well-meaning but uninformed employees move plants about, deciding that they are in the way or would look better somewhere else. This poses a problem because when a plant is moved and its light intensity changes, it must adapt to the new light level as best as it can, whether the intensity is lower or higher than its previous location. Plants moved to significantly lower light intensities exhibit yellowing leaves that may drop. Their new growth is spindly and pale in color, spaced farther apart than mature leaves, and is smaller in size. This is particularly visible in vining-type plants such as *Epipremnum* or *Philodendron scandens*.

TABLE 10-2: Available Light and Suitable Indoor Plant Genera

Low Light Intensity (50–100 Footcandles)	Medium Light Intensity (100–250 Footcandles)	High Light Intensity (250 and higher Footcandles)
Aglaonema	*Aglaonema* (*Variegated cultivars*)	*Adonidia*
Aspidistra	*Araucaria*	*Aechmea*
Chamaedorea	*Asparagus*	*Aeschynanthus*
Dracaena fragrans	*Brassaia*	*Aphelandra*
Dracaena reflexa	*Chamaedorea*	*Araucaria*
Epipremnum	*Chlorophytum*	*Beaucarnea*
Howea	*Cissus*	*Begonia*
Philodendron	*Cryptanthus*	*Caryota*
Rhapis	*Dieffenbachia*	*Chlorophytum*
Sansevieria	*Dracaena*	*Chrysalidocarpus*
Schefflera	*Dracaena fragrans* 'Massangeana'	*Cissus*
	Ficus	*Codiaeum*
	Hedera	*Coffea*
	Hoya	*Cordyline*
	Maranta	*Crassula*
	Nephrolepis	*Cycas*
	Peperomia	*Dizygotheca*
	Philodendron	*Dracaena*
	Pilea	*Euphorbia*
	Pittosporum	*Ficus benjamina*
	Polyscias	*Nolina recurvata*
	Spathiphyllum	*Phoenix*
	Syngonium	*Podocarpus*
	Zamioculcas	*Polyscias*
	Zebrina	*Zygocactus*

Delmar Cengage Learning

Plants moved to significantly higher light intensities could suffer, with new leaves being stunted and small, bleached, or even burned (due to heat injury). Only knowledgeable professionals should move plants from one location to another. Maintenance contracts should stipulate that if plants are moved by anyone other than an authorized interiorscape maintenance company member, they will not be guaranteed for replacement.

Light Duration

Light duration deals with the amount of time plants are exposed to light. Scientists refer to this as **photoperiod**. In nature, plants photosynthesize in sunlight, but during a cloudy day or at night, light duration ceases. An interior space with no windows provides light only when artificial lights are engaged. When the lights are off, the source is gone whether for the day, the weekend, or a staff vacation. Many indoor plants are highly tolerant of light duration fluctuation because they compensate with lower respiration rates (food use) in relation to photosynthetic rates (food production). Examples include dark green *Aglaonema, Dracaena,* and *Sansevieria* (Figure 10-10) varieties.

FIGURE 10-10: The tough, yet elegant *Sansevieria*.

Most indoor plants do best when exposed to light for 12 hours per day. They are able to maintain healthy leaves and overall form, thus staying good-looking for many months, even years. Light duration is fundamental in maintaining cost-effective plant installations.

Photoperiod has an effect on plant flowering, the reproductive phase of a plant's life. Most plants are day-neutral, which means that they are able to go into a flowering phase no matter how long or short the day length or photoperiod. There are indoor plants that are greatly affected by photoperiod, which is of more importance to growers than to interiorscapers.

Some plants that bloom in the summer rely on photoperiods longer than 12 hours. This makes sense because the longest day of the year is the summer solstice, occurring around June 21 in the Northern Hemisphere. Termed **long-day plants**, specimens such as *Calceolaria, Cineraria,* and *Hibiscus* need long days to trigger blooms. These plants are available from growers in the spring and summer months and are used to create focal points or extend a theme in an interior setting.

Some interior plants that bloom in the fall or winter require short days in order to reach their flowering phase. They need nights longer than 12 hours; even the slightest interruption of light can throw off their flowering. Because holidays of commercial importance occur during the fall and winter, interiorscapers often use **short-day plants** supplied by growers for seasonal displays. Examples include *Euphorbia pulcherrima,* Poinsettia; *Dendranthema grandiflora*, Chrysanthemum; and *Kalanchoe blossfeldiana.*

Shorter time frames may be compensated for with slightly higher light intensities. Spaces with weaker light intensities should leave lights burning for longer periods of time, from 12 to 18 hours (Figure 10-11). Plants need light to survive, and it is possible that they can be maintained in spaces with nothing but artificial light. Offices and retail spaces may be completely devoid of the full-spectral colors offered by natural light. Plants classified in low light categories may be maintained for extended periods, perhaps even years, in artificial light.

FIGURE 10-11: Some office spaces leave lights burning up to 24 hours per day. Longer duration of light compensates for low light intensities; thus, foliage plants remain healthy in appearance.

Light Quality

The sun is a burning star. It is the original light source for Earth. Little wonder that plant species evolved to vote it their favorite light source for the past 500 million years. The light from the sun is full of colors, and each of these colors aids in specific responses within plants.

Red and far-red light initiate flowering and stem elongation and are needed in photosynthesis. Anthocyanic (red) pigments, which give color to variegated plants such as *Codiaeum variegatum*, Croton, or *Maranta leuconeura*, Prayer Plant, rely on red and far-red light to trigger these bright leaf colorations. Blue light works with red to maximize the effects of photosynthesis. It is also responsible for phototropic response. Yellow-green light is bounced off of leaves. That is why we see it and it is not used as much by a plant. Thus, if a plant is grown in a yellow-green light, it will not be robust. Plants grown in any singular spectral color will not be healthy over time because other necessary colors are missing. This is generally not a problem, though, because people who keep plants live and work in environments using white light rather than singular colors. White light is the best artificial light.

Incandescent light bulbs are not as efficient as fluorescent bulbs in terms of radiation of usable light by plants. They are efficient in radiating red light, but not as much blue light. Compact fluorescent light bulbs are more energy-efficient than incandescent bulbs and are widespread in use (Figure 10-12).

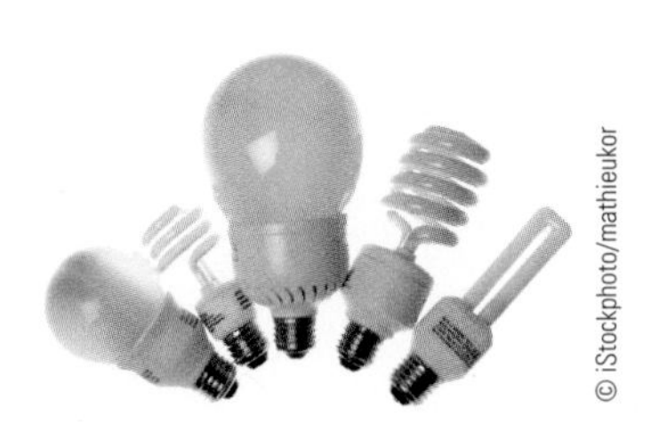

FIGURE 10-12: Compact fluorescent bulbs are manufactured in a variety of shapes, intensities, and colors.

All types of artificial lights can produce usable light that can sustain the maintenance of interior plants.

SUMMARY

Plants cannot live without light. If it is removed, the energy source from which they are able to manufacture their own food is missing and they will die. For the interiorscape technician, the availability of light for an installed plant makes the difference for how long that plant will keep its aesthetics.

The process of photosynthesis ensures the production of carbohydrates necessary for a plant to carry on its life processes. Through respiration, the carbohydrates made in photosynthesis are used to aid the plant to carry on the processes of life and growth. Photosynthesis rates must exceed respiration. Even though low indoor light levels can only sustain slow production of leaves and shoots, such situations can still offer the benefit of maintaining beautiful plants indoors.

CHAPTER 11

Water and Soil

INTRODUCTION

In photosynthesis, plants convert carbon dioxide and water to sugar in the presence of light. Once sugar is made, it must be transported throughout the plant for conversion into energy or for storage in the roots. In order for leaves to be the houses for photosynthesis factories, they must be held erect or **turgid**, and water helps to puff up cells so that this work can be done. In a way, plants are like straws that take up water from the ground through roots and disperse it into the atmosphere through stomata in the process of transpiration. In high light and warm conditions, plants are naturally able to keep themselves cool through transpiration, drawing water from the soil through root hairs (Figure 11-1), and then transporting that water through the stems to the leaves. As water molecules move from the leaves through stomata, they carry heat energy away from the plant. This is known as evaporative cooling. The way the human body sweats in order to cool itself is a similar phenomenon.

Water Flow

Water always moves from areas of high water concentration to areas of low water concentration. Since root hair cells (Figure 11-2) contain organelles and dissolved solids, water moves into them freely from the soil when the concentration of water is greater in the soil than in root hair cells. This **diffusion** of water through the cellular membranes is called **osmosis**. Water also moves throughout the plant because as it evaporates from leaf surfaces though **stomata**, its natural tendency to stick to itself causes it to move from cell to cell and outside the plant. When accomplished indoors, these activities help to improve indoor air quality that may be at levels that are too dry to keep people healthy. The moisture vapor that surrounds a plant also helps to cool the plant from heat. The vast amount of a plant's weight is water.

When a plant wilts, its stomata close in order to conserve precious moisture. When stomata close, respiration slows because carbon dioxide levels are low. This leads to a decrease in photosynthesis levels.

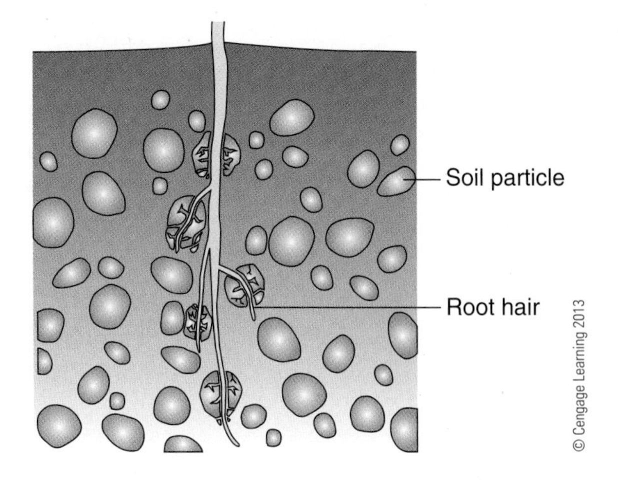

FIGURE 11-1: Root hairs within the soil.

Root Hairs

As tiny root hairs take in water, the soil pore spaces near the roots that have lost water molecules fill with air. Oxygen, carbon dioxide, and other gasses are literally pulled into those spaces. Root hairs need oxygen for transpiration because this is an area of great cellular division and growth. If the medium were waterlogged, carbon dioxide would build up in what should be air-filled pore space, thus limiting respiration. Root hair cells, and then tissues, would eventually die. The next stage would be the onset of rot where soil microbes would break down and decompose dead tissues. Ultimately, this will result in a dead plant. The deadly combination of using a medium that drains poorly and overwatering are common problems associated with the culture of plants in the interior.

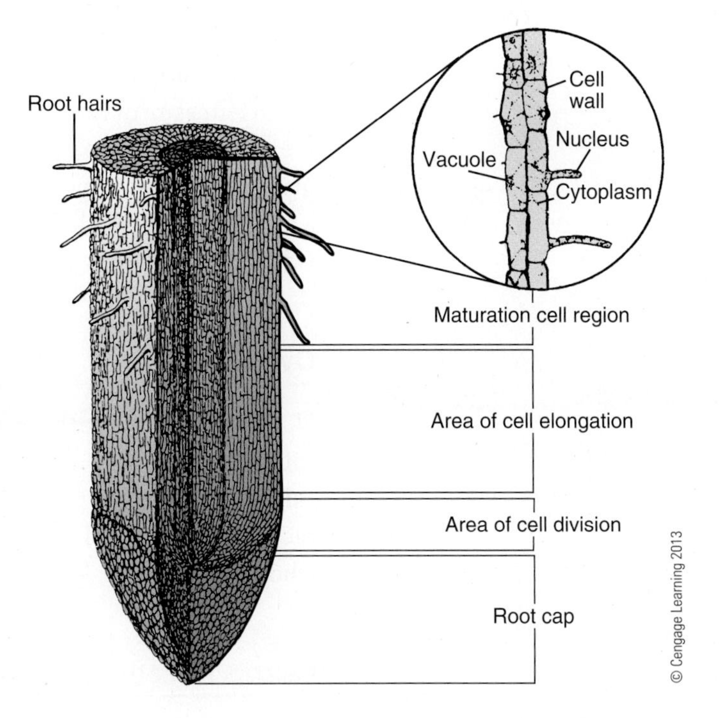

FIGURE 11-2: Root hairs are produced by the millions each day on actively growing plants.

Root/Shoot Ratio

A plant does not die just because some roots die. The plant is able to react efficiently with such changes in order to make the best of its situation. Since there is less water and nutrient uptake, the plant will shut down some of its photosynthesis factories, thus dropping some leaves. Depending on the severity of the problem, it may even kill off a branch or two. On the other hand, if a plant is healthy and in a growth situation with plenty of available light, it produces more shoots and leaves. Because more water is needed and more photosynthate is manufactured, more root growth occurs. This relationship is known as the **root/shoot ratio**.

Watering Plants

It seems simple, but balancing the amount of water plants receive is something of a tricky matter. With experience, and some errors along the way, an interiorscaper learns how to sustain the life of indoor plants with just the right amount and frequency of watering. This is one of the reasons why it is good to keep plants in your own home so that you can learn from practice. Pre-professional work opportunities such as internships and cooperative experiences provide additional learning. It is much better to drown an inexpensive *Areca* palm purchased for your apartment than a regal, slow-growing *Howea* in your employer's best hotel account. When you make a mistake, and we all do, it is best to make it in the least costly way. Plant watering techniques are explained in Chapter 15.

Overwatering

No life can occur without water. On the other hand, too much of a good thing is a bad thing. Some people kill plants easily by over watering. This can happen in numerous settings both in the home by plant lovers who indirectly become plant killers and in the workplace by interiorscape technicians who need additional training.

Factors Affecting Watering Frequency

Frequency of plant watering depends on many different factors that are dynamic, changing over time. They can all be overcome and interiorscape technicians can master the intricacies of plant watering challenges, making the employee valuable to employers and clients alike.

Media

FIGURE 11-3: This *Homolomena* can still be revived, but its highly water-stressed appearance reflects poorly on the plant maintenance crew.

Many factors affect a plant's watering needs. Factors include the age of plant, whether a plant is root bound, or if the root system is new with plenty of available media. The more total dissolved solids in the medium, the more damage dryness will cause to the plant. This is true for older, established plants where the soil environment has collected fertilizer salts and dissolved solids from water supplies over long periods of time. It is also true with new plants having high post-production fertilizer residues and not receiving the leach watering they need. If a technician is using a water meter, high levels of soluble salts will trigger meters to register the plants as "dry" even though the medium is moist.

Media age over time as organic material contents break down through microbial action. As time goes by, there are fewer air-filled spaces and more water-filled spaces. Smaller-sized potted plants in the 4- and 6-inch pot diameter categories dry out very quickly, resulting in a water-stressed plant (Figure 11-3). They can be nearly impossible to keep watered and, when used, should be kept in subirrigation systems. Larger grow-pot diameters generally require less watering due to the mass of soil within them.

Amendments used in potted plant media greatly influence how often a plant must be watered. Media high in peat can be a challenge to hydrate if allowed to dry. Sandy media are well drained, requiring more frequent watering than media lower in sand content when used to grow tropical plants like *Spathiphyllum* or *Dieffenbachia*. Of course, good drainage works in favor of cacti and succulents but may aid in death by dryness for maidenhair ferns or other moist-soil-loving plants.

Temperature

Warmer indoor environments cause soil moisture to evaporate more quickly and for plants to respire at higher rates, thus needing more frequent watering than similar plants in slightly cooler indoor environments. Increased light intensities from the sun can create heat, so technicians should be cognizant of these changes throughout the seasons. Foliage plants placed in drafty areas, even in the slightest of air currents, may need more frequent watering to compensate for the water that is quickly removed. Consider this when placing plants near exterior doors or within 10 feet of heating/air-conditioning ducts.

FIGURE 11-4: Plastic pots hold moisture longer than clay and are lighter in weight though not as decorative. Plastic pots are often referred to as grow pots and are concealed within more decorative containers.

Containers

Terra cotta containers are porous, allowing for transfer of moisture, but most indoor plants are potted in plastic containers, which hold moisture (Figure 11-4). Interiorscapers and growers rarely directly place plants in nonglazed, clay pots. They present great challenges because media and plant roots within them dry out so quickly. Often, plastic pots are concealed within clay pots with a decorative covering of mulch to conceal the grow pot. The only downside to this is that the terra cotta display pot does not receive the beneficial effects of being used, the months of soaking and drying coupled with the dispersion of soluble salts. Container designers and manufacturers artificially weather new clay containers with paints and textures to give containers an appealing, "Old World" or used-in-the-garden appearance. The placement of mulches on soil surfaces also helps to conserve water.

Plants in small pots require watering much more frequently than do those in larger pots. A large pot holds a greater volume of soil; thus, there would be a greater number of pore spaces. Using small plants in an interiorscape is problematic because they will frequently exhibit water stress and require more intensive maintenance schedules. Sometimes, new interiorscape designers will tend to utilize many small plants in an installation because they are accustomed to working with minor projects and budgets. These potted plants will dry out quickly and, due to the nature of soils with high levels of organic content, will be difficult to saturate. Upon watering the dry ball of soil, moisture will run down the space between the contracted soil ball and the inner wall of the pot, then stream below. The soil will not be saturated and soil pores will remain air-filled with no chance of becoming water-filled. The plant becomes stressed and dies quickly if the problem is not remedied.

A better choice in practice is to use a few larger plants rather than several small plants in the 6-, 8-, or 10-inch diameter pot sizes. Larger specimens cost more initially, but save the interiorscaper and client replacement expense over time. A larger plant has dramatic visual impact.

Water Quality

We might think of water as a basic thing that plants need, but we might not think too much about what is in the water that can affect a plant's health. Water quality refers to the components of the water. Plants are affected by soluble salts, hardness, and pH of water.

Soluble Salts

When we read about soluble salts, we may think only of table salt, sodium chloride, but more accurately, soluble salts are all types of minerals dissolved in water. They are so minute that they are measured in parts per million, but it does not take a large measure of them to make a big difference in the health of a plant collection. It is good to know that, in general, most municipal water sources are fine to use on plants. The exceptions to this are with plants that are sensitive to fluoridation such as *Dracaena fragrans* 'Janet Craig' and *Chlorophytum* Spider Plant. If these or other plants with long, pointed leaves exhibit tip burn, the **necrotic** or dead portion of the plant tissue can be trimmed with clean scissors (Figure 11-5). If such plants lose their beauty too quickly, they should be replaced with plants that are not sensitive to fluoride additives saving the interiorscaper maintenance and replacement problems.

FIGURE 11-5: Trimming is a regular part of plant maintenance. Strive to shape the leaf so it looks similar to healthy leaves.

Plant fertilizers contain minerals dissolved in water. Residual minerals can collect in soil or on soil or pot surfaces over time. As these salts concentrate over time, they can reach levels that are toxic to plants and action must be taken. To slow down the rate of salt build-up, the classical watering technique referred to as **leaching**, where water that carries excess salts flows through drainage holes into saucers that are then emptied, should be employed. A more involved step is to remove the plant from its pot, replace looser soil with new media, and repot in a scrubbed, sanitized pot. In this way, salty soils are removed and roots can be reestablished in the new medium. The brief amount of time and effort spent in repotting an older plant is well worth the effort.

Even without added fertilizer, salts already present in water concentrate in soil over time. The best way to know the exact concentration of a medium's soluble salts is through a soil test, available from state soil testing laboratories. Contact information for such labs is available through county extension office web sites. There is a big difference between soluble salt tolerances in outdoor or field crop soils and indoor, interiorscape soils. It is important always to identify soil samples as originating from an interiorscape.

Water Hardness

Some water supplies are **softened**, meaning that calcium and magnesium are replaced with sodium. This is accomplished with some water supplies in order to make them better for human use. Unfortunately, the greater concentration of sodium is harmful to the water balance needed by root hairs. Plants would retain their health with use of hard water; therefore, the water should be used before it is run through the softener apparatus.

pH

pH is the term given to the measure of hydrogen ions in solution. It is measured on a scale of 1 to 14 (Figure 11-6). Pure water has a pH of 7.0. Any measures below this point are increasingly acidic, and measures above it are alkaline.

FIGURE 11-6: pH values less than 7.0 are acidic, while those above it are alkaline. A pH of 7.0 is neutral.

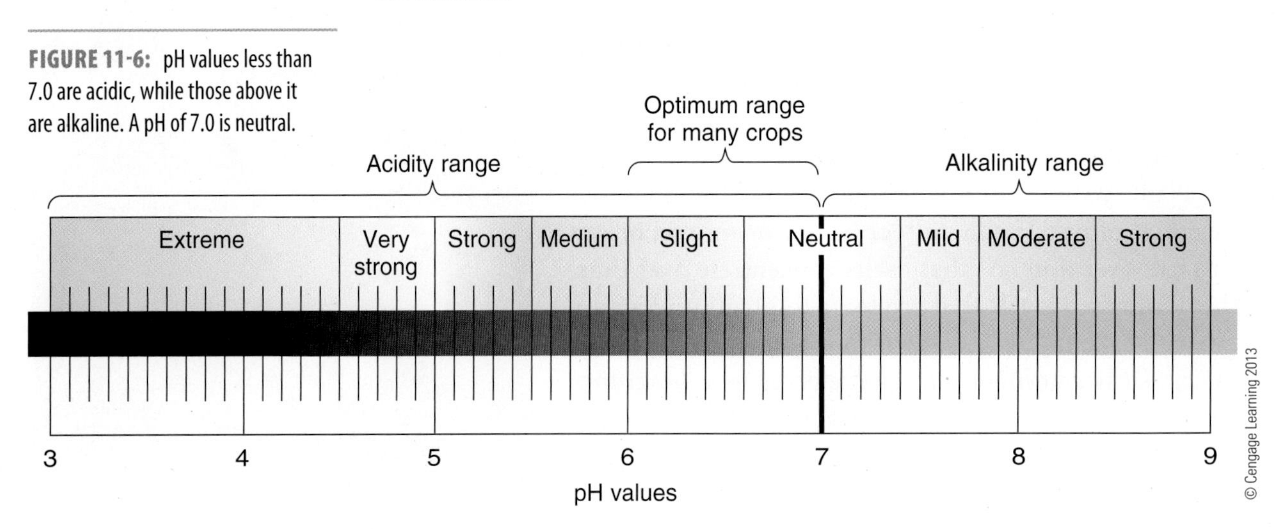

While indoor plants respond to water that is slightly low in pH (acidic), most municipal water supplies are slightly high (alkaline). This difference does not have a great effect on the health of indoor plants.

Interiorscapers should keep in mind that water moves from areas of high water concentration to areas of low water concentration. Root hair cells necessarily have a concentration of solids, including dissolved minerals. In order to keep the proper directional flow of water into root hairs, the greater concentration of solids should remain in root hair cells rather than in the soil. Reversal of this directional flow due to buildup of soluble salts in the soil results in necrosis of cells, tissues, and ultimately the entire plant organism.

Moisture Meters

Technicians using moisture meters must take into consideration the age of the plants' medium. These meters do work to measure moisture, but in an indirect way. They do not measure an amount of water, but the amount of electrical conductivity present in a medium. Older soils have had more time to collect soluble salts that can have misleading effects on moisture meter readings. The more total dissolved salts in a medium, the higher the conductivity of the medium. Therefore, an older planting that is dry and in need of water could provide a moisture meter reading of wet. Conversely, distilled water with nearly all dissolved salts removed can give a reading of dry. Infrequent use of a moisture meter has another positive effect because driving the probe into the soil can aid aeration.

Temperature

Damage to plant materials is caused by extremes in temperature and can be characterized in two different ways. An acute exposure, meaning that the damage is severe and sharp, reaches a crisis stage quickly and could happen when a tropical houseplant is carried outdoors on a cold, winter day, for example. Chronic exposure to cold or heat is differentiated as being prolonged or lingering. Office building vestibules may not be the best place for a live plant when there is regular exposure to hot summer air or cold winter blasts. Tropical plants are tolerant of temperatures down to about 40 to 45°F (4 to 7°C) for short periods of time. Exposing a tropical plant to temperatures lower than this for longer than a few minutes can result in rapid decline or death of all or part of the plant. It is very important that when plants are delivered to installation sites, they are wrapped to protect them from

FIGURE 11-7: The effect of chilling injury on *Dracaena*.

FIGURE 11-8: Freezing injury exhibited on a *Stromanthe sanguinea* leaf.

exposure to cold. Similarly, they must be protected from extremely warm temperatures. Care must be taken in transport of tropical specimens from production areas to markets. Shippers should avoid temperature extremes in long-distance shipping.

An acute symptom of temperature extremes results in downturned leaves or leaves that appear darkened. Areas of the leaves may have a "water-soaked" appearance, as if there is a slight transparency in spots (Figure 11-7). Freezing injury is more severe because water within the plant cells has solidified (Figure 11-8). Ice crystal formation pierces through cell walls causing the contents to leak or **lyse**, thus killing cells and tissue.

It is important to keep temperatures relatively constant when transporting and displaying indoor plants. Keep in mind the native habitat of plant varieties when transferring them to an interiorscape site. Communicate this information to horticulture technicians and clients.

SOIL MUST

- Provide for root anchorage
- Offer water-filled pore space for root life
- Offer air-filled pore space for root life
- Offer minerals needed for survival and growth

At a basic level, soil provides four main purposes for plant life: anchorage, nutrition, water, and oxygen.

Soils

The horticulturally uninitiated may regard soil in a potted plant as "dirt," muddy stuff of some unknown importance that helps a plant grow. Those in the know understand the root environment of plants is a place of active exchange, growth, and interaction. Although seemingly unglamorous, if it were not for the root environment of indoor plants, there would be no flower/foliage spectacle at all.

Anchorage

In order to remain upright, a plant must have something to hold on to. Roots are able to grow within the spaces of soil particles, lodging new

TABLE 11-1: 16 Elements Needed by Plants for Survival

Nonmineral		Primary		Secondary Mineral		Micronutrients	
Name	**Symbol**	**Name**	**Symbol**	**Name**	**Symbol**	**Name**	**Symbol**
Carbon	C	Nitrogen	N	Calcium	Ca	Boron	B
Hydrogen	H	Phosphorus	P	Magnesium	Mg	Chlorine	Cl
Oxygen	O	Potassium	K	Sulfur	S	Copper	Cu
Iron	Fe	Manganese	Mn	Molybdenum	Mo	Zinc	Zn

growth within so that the structures above the soil are able to develop and orient toward available light. Occurring in tandem, the photosynthetic factories of leaves can kick into higher production due to work being done in the soil.

Good anchorage is important to stability and therefore aesthetic balance of an indoor plant. Poorly rooted plants lack in visual and sometimes physical balance and are not appropriate for professional plant displays.

Nutrition

There are 16 essential elemental nutrients necessary for plant survival (Table 11-1). Think about all of the elements present on a periodic chart of the elements; there are well over 100, some of which are artificially made and not part of the Earth's surface layer. As horticulturists, we can think of those needed by plants as the "sweet 16." Because they are elements, they cannot be broken down into smaller quantities.

Elements exist in the form of atoms. Atoms are made up of a positively charged nucleus at its core and one or more negatively charged electrons that surround the nucleus. Because the number of electrons is equal to the positively charged nucleus, atoms are neutral. In the soil environment, atoms can change by losing or gaining electrons, thus having a net negative or positive charge. An ion is born.

Positively charged ions are termed **cations** and negatively charged ions are termed **anions**. Electrical charges are shown when designating ions. For example, some positively charged ions are H^+, K^+, and Ca^{2+}. Negatively charged ions are NO_3^-, Cl^-, and SO_4^{2-}. Most cations are a single element, but most anions are combinations of elements. Soil colloids, the majority of these being clay and humus particles, are negatively charged, so positively charged cations are attracted to them while anions are repelled (Figure 11-9). Because they are free and unattached, anions are more easily removed by water moving through the soil.

Hydrogen ions from the root replace cations present in the soil solution. The process where hydrogen ions from the root hairs replace cations attracted

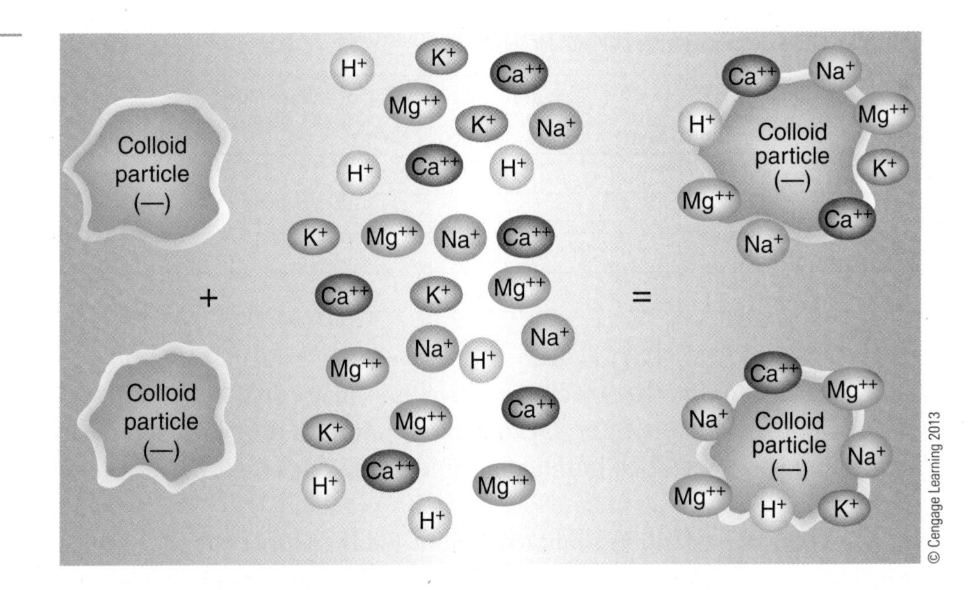

FIGURE 11-9: Positively charged nutrient cations are attracted to negatively charged soil colloids.

to soil colloids is called **cation exchange**. This exchange occurs when positively charged nutrients are released by soil colloids and positively charged nutrients are absorbed by roots.

Interiorscape plants may exhibit symptoms of certain nutrient deficiencies. Table 11-2 provides a list of the 16 essential nutrients, what they aid in plant growth and development, and symptomatic appearance in plants deficient of the particular element.

Water and Mineral Movement

Ions move into the roots and continue to concentrate in cells, thus aiding in water movement with them. Recall that water moves from areas of high water concentration to areas of low water concentration. The flow of water into the cells provides enough pressure to move water and minerals into the shoots and leaves of plants. Eventually, this pressure causes guttation, with the resultant small drops of water seen on the tips of leaves (Figure 11-10).

FIGURE 11-10: Indoor plants such as *Epipremnum*, *Dieffenbachia*, *Ficus*, and many more may exhibit guttation as water balances within the plant.

A second mechanism at work aiding moisture and mineral movement in plants is evaporation through stomatal pores. As water evaporates from leaves, it pulls additional water molecules through xylem tissue due to **water tension**.

TABLE 11-2: Functions of Nutrients in Plants and Their Deficiency Symptoms

Nutrient	Function	Deficiency symptoms
Nitrogen	Promotes rapid growth, chlorophyll formation, synthesis of amino acids and proteins	Stunted growth, yellow lower leaves, spindly stalks, pale green color
Phosphorus	Stimulates root growth, aids seed formation, used in photosynthesis and respiration	Purplish color in lower leaves and stems; dead spots on leaves and fruits
Potassium	Increases vigor, disease resistance, stalk strength, and seed quality	Scorching or browning of leaf margins on lower leaves, weak stalks
Calcium	Constituent of cell walls; aids cell division	Deformed or dead terminal leaves, pale green color
Magnesium	Component of chlorophyll, enzymes, and vitamins; aids nutrient uptake	Interveinal yellowing (chlorosis) of lower leaves
Sulfur	Essential in amino acids, vitamins; gives green color	Yellow upper leaves, stunted growth
Boron	Important to flowering, fruiting, and cell division	Terminal buds die; thick, brittle upper leaves with curling
Copper	Component of enzymes; chlorophyll synthesis and respiration	Terminal buds and leaves die, blue-green color
Chlorine	Not well defined; aids in root and shoot growth	Wilting, chlorotic leaves
Iron	Catalyst in chlorophyll formation; component of enzymes	Interveinal chlorosis of upper leaves
Manganese	Chlorophyll synthesis	Dark green leaf veins, interveinal chlorosis
Molybdenum	Aids nitrogen fixation and protein synthesis	Similar to nitrogen
Zinc	Needed for auxin and starch formation	Interveinal chlorosis of upper leaves
Carbon	Component of most plant compounds	
Hydrogen	Component of most plant compounds	
Oxygen	Component of most plant compounds	

Air-Filled and Water-Filled Pore Spaces

There is a seesaw balance between watering and drying of plant media. In other words, after a plant is watered, technicians must give the soil environment time so that the water-filled pore spaces may empty, through evaporation, drainage, or uptake by root hairs. This may take days or weeks to happen. Root hairs need air as well as water to function, so the pore spaces need time to empty before being filled again with water.

Moisture in the media is important because minerals are solids and are only available to plants when they are mixed in water. Since roots can only take up these solids when they are mixed in water, we should consider the soil environment a solution rather than a solid.

Roots cannot remain in a waterlogged soil environment, however. Roots are living, respiring plant parts, and in order to remain so, they need oxygen in order for respiration to occur. Roots growing in soil cannot remain under water because they will not receive oxygen and will not perform nutrient uptake. In turn, as roots respire, they give off carbon dioxide, so there must be air-filled pore space in media to allow for gas exchange. Air-filled pore space is the volume of the growing media that remains filled with air after a thorough watering followed by one day of drainage. For indoor, potted plants, about 20 percent of the soil environment should be air-filled pore space and about 40 percent should be water-filled. In order to achieve this, the media must have low, if any, levels of mineral soil.

Field Soil

Using field soil, also known as mineral soil, for indoor plants is not recommended. It must be pasteurized in order to kill pathogens and weed seeds. Because field soil is high in clay particles, it is not well drained in containers, having high water-filled to air-filled pore space ratios. Clay is very high in surface area and holds water molecules strongly. It adsorbs and retains nutrients that can be used by plants.

Consider a beautiful exterior garden at someone's home. This property may have lush, healthy plants and flowers, partly because of the excellent soil with very good drainage. The same soil removed and added to a pot may result in poorly performing plants because of what is termed the **container effect**.

Field soil is in a very long column in its natural surroundings. Gravity aids in pulling water through the soil for long distances (Figure 11-11). The water table, where well-drained soil and water-saturated soil interface, may be 10 to 20 feet or more below the soil's surface. This same heavy soil's water table will be proportionally higher in a pot, and thus closer to roots.

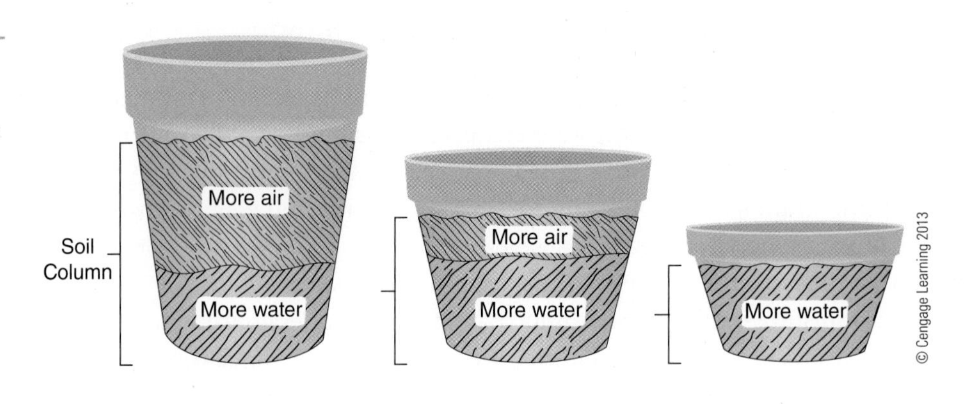

FIGURE 11-11: Depth of the soil column has much to do with drainage. Gravity allows deep soil columns to drain while shallow columns remain moist. Plants requiring well-drained media can perform better in tall pots while others with fibrous, moisture-loving roots may do better in shallow pots.

Field soil's ability to hold water, along with a higher water table, could cause **anaerobic** conditions for roots, leading to their decline.

Designed Soil

Countless research studies have proven plants can be grown successfully in numerous substrates including ground tires, insulation, straw, and more. As long as plant roots have moisture and nutrients, they are able to supply the plant with what is needed for survival.

Potted indoor plants require soils that have been altered to compensate for the container effect. Coarse amendments like sand and perlite are added to increase pore sizes between the particles. Care should be taken to monitor indoor plant installations over time. Over a period of years, artificial media will lose air-filled pore spaces and increase water-filled spaces. This compaction will be detrimental to plant health because roots need the gas exchange performed in air-filled pores. If plants exhibit symptoms associated with overwatering and compacted soil is suspected, repot the plants with new, fresh media.

In both of these instances, plants can have a wilted appearance, and over a longer period of time, leaf yellowing will occur. The symptoms of plants suffering from too much water or not enough water in the soil are the same.

There are great differences between field or mineral soil and manufactured or artificial media. Artificial media can be amended field soil, that is, pasteurized mineral soil with added elements such as peat moss and styrene beads. Artificial soil may not contain any field soil at all. In the design of artificial media, manufacturers know they are producing media for indoor plants. Indoor plants are often grown in smaller pots with less volume and depth and are prone to drying out more quickly. Mineral soil contains less than 20% organic matter. If a media contains more than 20% organic matter, it is classified as an organic soil.

Organic Material

Organic materials provide energy for soil microbes and nutrients for plants. It also helps a soil retain moisture as well as promotes air-filled pore space. There is little wonder that better indoor plant media mixes are high in organic material.

Organic material offers microbes food in two phases. The first is shorter-lived and consists of harvested plant materials of new tissue that is not decomposed. At this stage, soil microbes actively break down plant material particles and tie up much of the available nitrogen in the soil so that it cannot be taken

FIGURE 11-12: Humus.

up by plants. Once on its way to decomposition, the second form is much more useful to horticulturists as a soil amendment: broken-down organic debris known as humus (Figure 11-12). At this point in the life of organic material, microbial activity is slower and nitrogen is more readily available to the plant. Composting allows for organic materials to be fully rotted, transitioning microbial activity from being nitrogen-needy to lower levels of activity so that the benefits of air, water, and nutrition can be derived by plant roots. Studies show that plant roots and soil microbes are able to "communicate" via the release of volatile organic compounds from each other. This relationship results in bioactive compounds that work against the proliferation of pathogens and the promotion of plant growth.

Amendments

Several amendments are used to create indoor plant media. Amendments may be organic or mineral in origin. Commercial mixes of various proportions of these amendments are available to suit the needs of growers, interiorscapers, and hobbyists. Commonly used materials that are found in potting mixes are discussed next.

Peat Moss

Peat moss is harvested from many different geographic locations, but is also becoming scarce. It is the product of marsh mosses that have been preserved under water, so it is a limited resource. It is often used with other amendments because of its hydration properties. Wet peat moss stays saturated and is not easily aerated. If allowed to dry, is becomes somewhat hydrophobic. This can be problematic with indoor plants using greater proportions of peat in the media mix. If allowed to dry out, it can be difficult for water to penetrate fully the upper crust of the soil. This crust is referred to as a hard-pan. The addition of other amendments with peat allows for water penetration and better aeration. Non-ionic wetting agents help peat mixes absorb water. (See Figure 11-13.)

Perlite

Perlite is a heat-treated product made from volcanic lava. It appears to be small, white pebbles and is able to provide aeration as well as water-holding properties to media.

© Brandon Blinkenberg/www.Shutterstock.com

FIGURE 11-13: Peat, perlite, and vermiculite.

Vermiculite

Vermiculite is the product of heat-treated mica from natural deposits. It is available in different sizes and has the appearance of miniature accordions that absorb water and add good aeration to soils. If compressed, vermiculite loses these effective properties.

Sand

Sand adds good drainage properties to soil mixes and this is its chief attribute (Figure 11-14). It does not hold nutrients. Succulent plant materials such as *Agave*, *Mammilaria* and other cacti, *Crassula,* and *Yucca* benefit from the good drainage provided by sand as a soil amendment.

Sphagnum Moss

Sphagnum moss is a dried bog plant and is sold in two forms: milled and unmilled (Figure 11-15). The milled form resembles rough sawdust. The unmilled form is coarse with plenty of pore space. Some plants are directly grown and sold in sphagnum moss such as orchids. Sphagnum moss contains a natural anti-fungal substance.

© Stocksnapper/www.Shutterstock.com

FIGURE 11-14: Sand in bulk prior to mixing.

FIGURE 11-15: Long-fibered sphagnum moss is used for air layering and cutting propagation, lining hanging baskets and planters outdoors. It is ground to a finer consistency when used as a soil amendment.

Once seeds are sown, milled sphagnum moss can be added in a thin layer on the top of the soil to prevent damping-off fungus from attacking young seedlings.

Soil pH

For most indoor plants, a soil pH of 5.5 to 6.5 is optimal. Where fluoridated water supplies could affect indoor plants, a pH of 6.5 to 7.0 is recommended. Consult with state water testing labs when operating a professional interiorscape business. Soil and water testing labs will help with recommendations, saving horticulturists from costly errors. Because of the major differences between field soil conditions and interior plant conditions, always identify test samples and general inquiries as relating to interior plants rather than exterior landscapes.

As an installation matures, the soil environment also changes. Along with compaction, the amount of total dissolved solids, also called soil salts, increases. This salinity is the result of dissolved fertilizer solids as well as solids present in tap water. A signal that media contains high levels of soluble salts can be seen where white crusts appear near the pot's drainage holes, and on edges of the soil surface in plastic pots. Salts can leach through porous clay pots (Figure 11-16).

Propagation Media

Interiorscapers do not propagate new plants often, but interior plant enthusiasts and production specialists do. Many people are interested in adding numbers to their plant collections or sharing new plants with others. It is possible to root new plants in many types of soils. The best choice would be a medium that is well drained to avoid pathogens. Mixes of peat and perlite or vermiculite alone provide anchorage for the cutting until new roots emerge, but they do not offer nutrients for a growing plant. A technique used at home is to root cuttings in water. It may take time for roots that have been cultured in water to transition to soil, but the process is usually successful.

FIGURE 11-16: Old clay pots with soluble salt deposits.

Mulch

Adding a topping to the soil surface of plants displayed in decorative containers adds much to the natural beauty of a plant. Mulches often hide nondecorative grow pots within jardinières and add a finish to larger plant installations, importantly helping to slow down evaporation from the soil, lessening the frequency of watering. They also aid in slowing down soil compaction from watering the plant on top of the soil because it would shield the initial impact of water and disperses it in a wider area. The moist environment at the top of the soil, just underneath the mulch layer, is richer in organic material being decomposed by soil microbes.

Many different forms of mulches are available through interior plantscape supply companies. Heat-treated bark chips do not harbor pathogens or weeds, unlike bark chips available at retail nurseries, which should not be used indoors. These types cost more but are well worth the price for being clear of fungi or insect eggs.

A type of popular mulch is made from wood fiber but has the appearance of Spanish moss (Figure 11-17). These types of products have an added flame retardant making them safer for public space use. Various mosses such as sheet moss, mood moss, and Spanish moss, not a true moss but from the genus *Tillandsia,* are sometimes used for single-specimen plants or for tabletop gardens. Other types of plant dressings include preserved Reindeer moss *(Cladonia)* in its natural, gray color or dyed. Reindeer moss is not a true moss, but a lichen. *Amphidium* or Bun moss is yet another moss offered through floral supply sources. As with any type of soil covering, sanitation is of the utmost importance so that pathogens, insects, or weeds are not introduced into an interiorscape. Fully dried and preserved materials can provide interest and unique design when combined with live plants, differentiating products in the marketplace.

FIGURE 11-17: Flame-retardant-treated wood fiber is commercially made and provides a mossy topdressing for plants.

GREEN TIP

Mulches and the technique of layering grow pots within decorative pots aid in reduced watering needs. This not only conserves water but saves maintenance trips.

Natural river stones provide a natural, artistic finish to plants, but they are heavy to transport to installation sites. Because they are loose, people may be prone to pick them up in high-traffic areas necessitating periodic replacement. Some companies have recreated stones from lightweight rubber and recycled poly resin, and some have even attached these to plastic netting so that they can be installed quickly and stay in place.

SUMMARY

Plant roots exist in an environment of organic and inorganic particles in water. Within this environment, plant roots take in water and nutrients while expanding and stabilizing the foundation for plant parts above the soil.

Plants need water for a myriad of reasons. Water is completely interactive with the plant and within the soil. Sometimes there is too much of it, but for professional interiorscapers, more thought and effort are spent on its conservation. Plants' need for water is one of the chief reasons why regular maintenance is so important. Without water availability in the right balance, interior plant installations become unsightly or are not able to exist.

The best way to learn how to take care of indoor plants, especially proper watering practices, is to do it yourself. If you have never kept an indoor plant, now is your chance. If you are taking a class in interior plants, more than likely you have grown dozens and dozens of plants over the years. Try some new varieties. Speak with other plant enthusiasts and professional horticulturists about what they are growing and what media they like to use. Good horticulture involves understanding not only the theory behind plant life and processes but also the practice of taking care of plants.

CHAPTER 12

Insect Pests

INTRODUCTION

Perhaps one of the biggest problems of keeping interior plants is that they may become infested with insects. If we ask interiorscapers what their greatest concerns are, insect infestations will probably not be at the top of the list, but it may be in the top five. Insect infestations are sort of like bad news; you know that you are going to experience it from time to time. The important thing is to manage it when it happens, preferably before the insects set in and cause problems.

Some insects can do substantial damage to plants while others are more of a nuisance because they fly about the spaces occupied by both plants and people. Paradoxically, those insects that are most difficult to control (mealybugs, scale) do not bother people who live or work among the plants because they do not fly.

When plants are infested, horticulturists must evaluate the entire situation. The value of the infested plant must be considered along with its proximity to other noninfested plants. The overall health of the entire collection is at stake when one or a few plants are found with evidence of bugs. Another factor is the importance of time. The longer that harmful insects are enabled to linger and reproduce, the more damage will be done, and the more they will spread. These reasons add up to the fact that plant techs should be good scouts for plant problems. They should inspect plants at every maintenance visit to identify problems in the earliest stages possible, looking at leaf undersides, petioles, stems, and soil surfaces. They should note anything that looks out of the ordinary and report the problem to a supervisor.

Quarantine

An important thing to learn about insects is that there will be fewer incidences of infestation if plants are quarantined prior to their introduction to an existing collection. A plant quarantine means that a plant is kept away from other plants, perhaps 30 feet or more, for a period of one to two weeks. Some experts suggest a month of quarantine. By this time, an infestation, even if initially minor, would be more readily seen. The most common way that an insect infestation enters a collection is through the introduction of one infested plant. Proper quarantining raises the issue of time management when ordering and receiving plant materials. This should be accomplished well-ahead of installation schedules. Installing plants within one or two days of receipt is risky for the well-being of the established interiorscape.

Actions

There are a series of ways of controlling insect populations, from the very simple to the specialized. A good interiorscape firm deals with insects in the least invasive way initially, increasing means of control more assertively if necessary.

KNOW THE TERMS

Insect pests affecting plants = *Infestation*

Disease affecting plant = *Infection*

Insect Removal

The least invasive means of controlling insect populations is to gently but firmly remove them. This can be accomplished with cotton swabs in order to reach within leaf axils where bugs live. With time and diligence, it is possible to lower insect populations this way. Some plant care enthusiasts who maintain plants at home add isopropyl alcohol, also known as plain rubbing alcohol, to water and use it to swab away insects. Professional interior plantscapers cannot legally follow this practice, however. Alcohol is not labeled as an interior plant pesticide. Using chemicals in ways inconsistent with label directions is a violation of federal law. These regulations are not in place to be burdensome, but to protect people from harm.

Plant Removal

Often, the best thing to do with a heavily infested plant is to discard it. This can be very difficult for someone who is psychologically attached to a houseplant. For instance, they may have maintained a plant that was given to express sympathy. The plant is much more than an accessory. Though overgrown or devoid of the attributes of an aesthetically pleasing plant, it is symbolic of the life and death of their loved one, an extension of their spirit. It is also a reminder of the relationship between the sender and the recipient. Interiorscapers, on the other hand, keep interior plants chiefly for aesthetic purposes. Plants are an important part of the interior design and are reflective of their client's status, style, and, if a commercial environment, **brand**.

Pesticides

Any persons working for an interiorscape company and planning to apply pesticides must successfully pass a pesticide applicator license examination. When it is deemed necessary to use chemical pesticides, interiorscapers should follow safe handling procedures. They should only use chemicals labeled for indoor use, as recommended by agriculture departments within the state where pesticides would be applied. Label recommendations must be followed precisely. Inaccurate mixing estimates can harm plant materials and, over time, build insect **resistance** to pesticides. Solutions should be mixed in small amounts so that no excess remains to be stored or disposed.

Insecticidal Soaps

Insecticidal soaps are effective, safe, and popular. They are specially formulated to remove the coatings and disrupt the interior membranes of insects. Concentrates are available on the market and it is often best to mix them with distilled water because minerals dissolved in tap water may render useless the long, fatty acid chains that make these insecticidal soaps effective.

Since insects must be covered with the solution, it is important to coat the areas of the plant where the insects are living (and hiding). An insecticidal soap application is not a one-time shot. Applications must be repeated over time. All areas should be coated with the solution, which should not be wiped off or rinsed from the plant.

FIGURE 12-1: Plant care technicians must be highly aware of any type of spills they have caused and clean them immediately. It is preferable that any type of spray application be accomplished off-site.

Oils

Horticultural oils are made from highly refined paraffinic crude oil and can be used safely on many indoor plants. They work by blocking insects' air intake. Neem oil is extracted from the seeds of the Neem plant (*Azadirachta indica*). It is characterized by a brownish color and garlicky smell. Neem oil reacts with insects' hormones causing them not to feed or breed.

A downside is that interiorscape techs must be careful with over-sprays, portions of the application that do not stay on plant parts, but cover the walls, floors, and furniture (Figure 12-1). Over-sprays can be hazardous to passers-by, causing them to slip and fall. Mists can discolor and ruin wallpaper, painted surfaces, and textiles. In addition to this, the application will cause lay people concern. No one wants to be around when the tech is "spraying for bugs."

Any type of spray-on control may require removal of plants from interiorscape sites to a facility where they can be treated. It may require application conducted in off-hours and during weekends. In any case, it requires good discretion and much care on behalf of the technician.

GREEN TIP

Terry and other soft cloths can be laundered and reused. Take care to have them in ample supply so that the same cloth is not used to clean several plants. This may vector insects and disease.

Systemic Insecticides

Other pesticides are applied as systemics, added to the plant's soil as solids, which will break down, or as liquids, mixed with water and poured onto the soil surface. Systemic pesticides are taken up by the roots and distributed throughout the plant's vascular system. A common systemic insecticide, **Imidacloprid**, affects insects' nervous systems causing them to die. When insects eat plant tissues or suck plant juices, they ingest the pesticide. Imidacloprid is harmful to beneficial insects, so great care should be taken when considering its use. In order to be good stewards of the planet, interiorscapers should always use the least-harmful means of dealing with pests. There are other types of pesticides registered for interiorscape use that control specific insects. It is always best to consult with your state extension service for up-to-date recommendations.

The Culprits

Many pests are associated with indoor plants. Some plants are susceptible to specific insect pests; for example, spider mites enjoy Ivy. Many of the common indoor plant insects feed upon a variety of plants. The following insects are the most common ones affecting indoor plants in offices, homes, and greenhouses.

Aphids

Look for aphids on new shoots. Plump-bodied, aphids have piercing stylet mouthpieces and suck plant juices, drying out plant cells by the thousands. They choose to feed on the softer parts of plants such as new shoots, flower buds, and undersides of leaves. Aphids can appear in varying colors, from greens to reds or yellows. When conditions seem to be changing, such as overdeveloping populations or when delicious, nutritious plant juices are drying, aphids can develop wings and fly away to better prospects.

Four secondary problems can occur from an aphid infestation. First, they continually move their stylets from cell to cell, not only sucking the life out of the plant, but also potentially spreading infection with their needle-nosed sticks. Besides being disease vectors, they excrete honeydew. Honeydew is concentrated with plant sugars and leaves a sticky residue on anything underneath the aphid. Add hundreds of aphids and you have a clear but sticky mess on clients' fine wooden tables, marble flooring, or luxury upholstery. At this stage, a client may think the plant is gently weeping some type of plant sap and may call the problem to the technician's attention. What can make matters worse is that the excrement does not stay clear for long but allows for the growth of Black Sooty Mold, a fungus that grows on top of honeydew. The mold makes honeydew more visible, but actually, the problem is from above, in the new growth regions of the plant. The fourth problem associated with an aphid infestation is that carbohydrate-rich honeydew attracts ants. Ants literally farm aphids on plant growth to harvest honeydew (Figure 12-2).

FIGURE 12-2: Ants tending to a flock of aphids.

If just a few aphids are present, they can be removed with a moistened cotton swab. A good, first-measure control if many insects are seen is to prune infested branch tips and gently but quickly remove the infested cut plant material to a bag and seal it. Larger infestations may warrant removing the plant from indoors and taking it to a shaded area where fine water mist blasts can be used to knock off most of the insects. The procedure should be repeated in a few days to knock off lingering pests.

Insecticidal soaps provide a good control for aphids. Some interiorscape companies create their own home remedies for the control of insects, using ingredients such as liquid hand soap mixed with water and sprayed periodically with no rinsing, in some cases monthly, on leaves and stems. Insecticidal soaps or ultrafine oils can be applied when the next step of control is necessary. Care should be taken to not apply any liquids to leaf surfaces in the presence of strong sunlight, either outdoors or indoors. Moisture droplets act as mini magnifying glasses and can burn leaf surfaces. Imidocloprid as a systemic insecticide can be applied if the previous controls are not working.

Fungus Gnats

Fungus gnats are black flies with long, delicate legs. They are not the worst pest that can infest indoor plants, but because they are able to fly about, they are an annoyance. Adults tend to fly around the soil surface and are attracted to moist, shady areas. The harm fungus gnats cause occurs during their larval stage where they feed on organic matter in the soil and damage roots. The injury they cause can provide an entry port for soil-borne disease. A way to identify fungus gnats is to place a wedge of raw potato in the soil. Fungus gnat larvae, if present (translucent with a black head), will feed on the potato. This works for houseplants and perhaps small interiorscape installations, but know that the wedges should be replaced every other day.

Heavily infested plants should be removed from the interiorscape. The best way to avoid fungus gnat infestation is to avoid overwatering, thus taking away a major component of their favored environment: moist soil. Insecticidal soaps or horticultural oil acts as a control and applicators should direct sprays to the soil surface.

Leafminers

Sometimes, ornamental plant material may arrive for interior display use and have unusual markings in the foliage. The cause of these "mini trails" through the interior of the leaf is the action of leafminer larvae. The larvae drop from their mines to the soil surface to pupate, and then become adult flies. Leafminer flies can be a nuisance and are similar in appearance to shore flies.

Damaged leaves should be removed and disposed of as soon as possible to prevent insects reaching adulthood and the generation of future generations. Leafminer adults are attracted to the color yellow, so yellow sticky cards are employed to help lower populations. Heavily infested plants should be removed and destroyed.

© iStockphoto/jeridu

FIGURE 12-3: Mealybug adults.

Mealybugs

An insect that is closely related to scale insects, mealybugs are persistent. They actively move at the crawler stage, when their bodies appear to be little white lines. At the adult stage, mealybugs are more oval-shaped and coated with a white, cottony, waxy covering that is waterproof (Figure 12-3). Adults produce honeydew and do not move. Look for them underneath leaves, in leaf junctures, and near roots. Ants farm mealybugs about and protect them from parasitic wasps, a beneficial insect. To learn more about beneficial insects, see "Biological Controls" later in this chapter.

Mealybugs can be removed manually with insecticide-dipped swabs. Take care to use many swabs rather than just a few when they are employed for insect removal. You do not want to transfer insects throughout a plant, and swabs are too inexpensive to make this mistake. Other safe and effective controls include horticultural oil. If these are not effective, Imidacloprid-containing insecticides are a third step.

Scale

Scale infestations can be tough to control because they have evolved to shed moisture due to the shell-like coating and helmet-like form of their bodies (Figure 12-4). Two basic divisions of scale can invade interiorscape plants. They are soft scales and hard scales. Table 12-1 illustrates some of the differences between these two types of scales.

Congregating on stems, petioles, and leaf-undersides, females lay from around 100 to 1000 eggs then die. Although dead, scales leave behind their helmet-like, hard covering in earthy colors like shades of brown and gray, which protects developing eggs. Because the covering is water-resistant, it is more challenging to penetrate it with insecticidal soap.

Try to remove scale manually with cotton swabs dipped in soapy water. If the infestation is too great, remove the plant and treat it with an insecticidal soap spray. The applications should be repeated according to label directions. Continue to remove scales, both dead and living, until they are gone. Insecticidal soap applications may be necessary for control, and multiple applications

FIGURE 12-4: Soft scale on a houseplant stem.

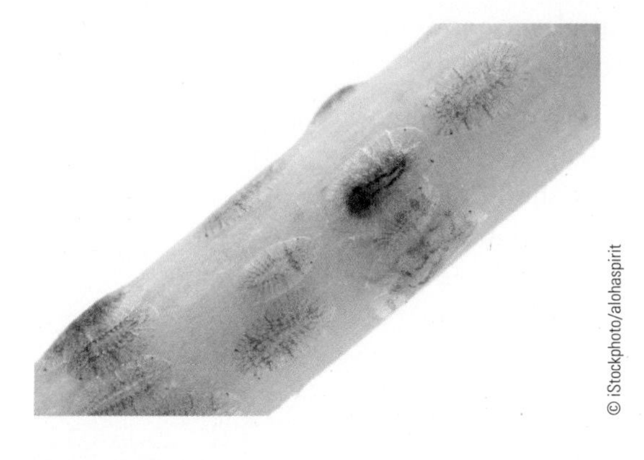
© iStockphoto/alohaspirit

TABLE 12-1: The Difference between Soft Scales and Hard Scales

Soft Scales	Hard Scales
Rounded and dome-shaped	Circular or elongated in shape
Excrete honeydew; problems associated with honeydew can occur	Non-producers of honeydew

over weeks of time may be needed, especially because of the scale's hard body coating. Imidocloprid-containing systemic insecticides can be quite helpful with large plants with scale infestations.

Shore Flies

Shore flies are considered nuisance pests indoors. Besides leaving small, black fecal deposits, they are able to fly about, thus distracting plantscape clients and their guests. A certain amount of this activity will indeed generate a phone call to your company to get rid of the little black flies. Shore flies resemble houseflies but are much smaller. They can be easily confused with fungus gnats, but shore flies are more efficient fliers (Table 12-2).

In order for shore flies to proliferate, they need moist soil surfaces. They do not feed on plant roots, but they do feed on algae located on media surfaces. If media is not constantly moist, algae will not grow and shore flies will not have interest. Shore flies do not find indoor environments like offices and malls the best place to live. They are more likely to be seen in greenhouses where moist air provides more growth that is algal.

Spider Mites

Two types of mites can cause damage on interior plants, broad mites and spider mites, though spider mites are more common and cause more damage. Broad mites cause disfiguration of new growth. Spider mites enjoy dry air; broad mites enjoy high humidity levels. Take care to inspect arriving plants used for new installations for stunted, disfigured growth. Remove any plant from an installation that displays this symptom. If problems persist, delve

TABLE 12-2: Look-Alikes: Shore Flies vs. Fungus Gnats

Shore Flies	Fungus Gnats
Nuisance pest	Nuisance pest
Better fliers	Long legs
Feed on algae on media surface	Cause root damage

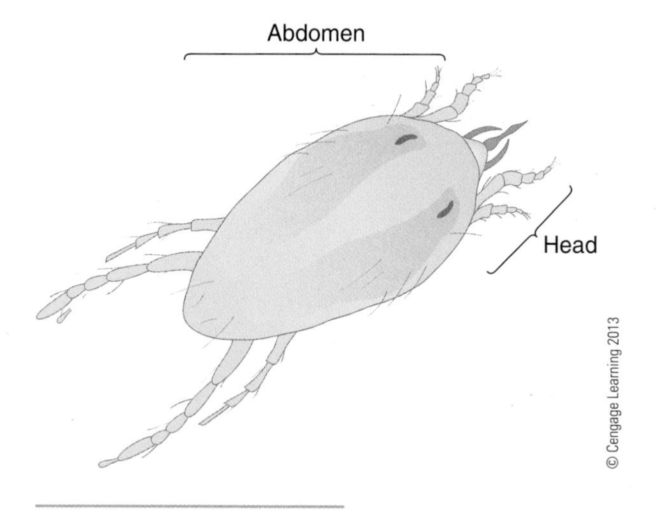

FIGURE 12-5: Mite anatomy.

FIGURE 12-6: Spider mite webs are a sign of infestation.

further to see if humidity levels can be adjusted at the site, depending on the mite

Perhaps spider mites are more problematic than broad mites are for interiorscapers because indoor environments are more likely to be warm and dry than very humid. Plantscapers who take on interiorscapes with these environmental conditions will face spider mites eventually. Infestations may occur in the dead of winter for northern interiorscapes or the heat of summer in the south.

Spider mites are not truly insects but spiders with piercing, sucking mouthparts (Figure 12-5). They congregate on leaf undersides and are impossible to see with the naked eye (Figure 12-6). One of the signs of a spider mite infestation is a slightly silvery cast to leaves, non-characteristic to an otherwise healthy, green-color plant. This is a result of cells dying, one by one, due to their contents being eaten by colony members. If very fine webbing is seen, the infestation is worse because it takes a longer period for the spiders to accumulate and make visible webbing. Spider mites can appear in numerous colors including greens, black, yellows, or reds.

Over-fertilization with nitrogen-containing fertilizers may hasten the growth and reproduction of spider mites. Keep plants away from drying airflows such as near air intakes or drafty entryways. Some houseplants just seem to attract mites more than others. *Hedera helix* and *Gardenia* seem to attract mites when under stress, especially dry soil.

Spider mites are hindered by soaps and oils, so periodic sprays with insecticidal soaps can aid keeping spider mites from gaining hold on leaves as will increasing humidity and keeping a plant clean. Dust removal is another benefit of using soap solutions. A simple way of increasing humidity is to set plants on trays of pebbles. Add water to the tray so that the water line is just below the point where the pebbles and the bottom of the pot touch.

The idea is not to set the plant in water, but to allow water vapor through evaporation to humidify the air.

Wash your hands when you work with mite-infested plants. They are very easily transferred throughout a collection and can be vectored by your hands and clothing.

Thrips

Thrips usually enter interiorscapes on color crops, such as *Cineraria*, *Cyclamen*, *Exacum*, *Begonia*, and chrysanthemums, introduced into larger plantings. This is significant because thrips are attracted to plants while they are in flower. Thrips have a rasping-style of eating where they use their mouthparts to scrape tissue surfaces and then suck plant juices. Sometimes the injury is in a linear form, as in the case of unfurling *Aglaonema* leaves.

Slender, with fringed wings, their color is normally light brown to black. Thrips live in dark, confined areas but feed on new growth with their piercing, sucking mouthparts. Foliage may appear silvery and stippled due to the way they feed on and kill plant cells. Peppery black specks on leaves are actually their excrement, another sign that a thrip infestation exists.

Since thrips, like spider mites, dislike humidity, increase relative humidity around plants. Always inspect new flowering plants for signs of thrips, and remove heavily infested plants and discard them so that they are not introduced into the interiorscape again. Occasional spraying of insecticidal soaps helps to counteract thrip populations. Greenhouse growers often use blue-colored sticky cards to attract adult thrips away from plants. They are attracted to blue more than yellow sticky cards, which are best for trapping whiteflies, winged aphids, leafminers, fungus gnats, and shoreflies. For testing purposes, sticky cards can be made with rigid plastic pieces cut to an appropriate size, about 2 to 3 inches across, and coated with petroleum jelly.

Whiteflies

Whiteflies, like thrips, hitchhike on flowering plants. Beware during the Christmas and spring seasons when introducing bright seasonal color that you are not sending along whiteflies to your otherwise pristine interiorscapes. For the most part, whiteflies do not endure in many indoor situations because the environmental conditions are not the best for them to thrive. They are more of a problem in greenhouses and conservatories where they can persist over time.

© iStockphoto/alohaspirit

FIGURE 12-7: Whitefly.

Whiteflies have piercing, sucking mouthparts. Their feeding on leaf undersides in newer growth results in stunted growth or leaf yellowing. They are easy to spot and, if not such a nuisance, are beautiful creatures due to their uniform white wings and bodies (Figure 12-7). Sometimes, with larger infestations, ants may appear along with whiteflies to farm them in order to harvest honeydew excretions. This activity, coupled with the fact that they are capable of flying about, makes them problematic to clients.

Adult whiteflies are attracted to the color yellow, so growers often place yellow cards coated with adhesive within crop areas to attract them. The standard procedure of looking over new plants closely will aid in catching a whitefly problem before infested plants are introduced indoors. Infested plants should be removed and all nearby plants should be scouted for problems. Occasional applications of insecticidal soap are recommended. Whiteflies are also controlled by Imidacloprid.

Pesticide Application

At least one employee of an interiorscape firm should be specialized in the identification and treatment of interior insect pests. This requires testing and passing a state-administered examination. Online training programs are readily available and include information, techniques, test preparation, and examinations. Most of the treatments recommended in this book for the control of indoor plant insects do not require pesticide applicator licensing, but as a company grows in size and takes on larger accounts, statistics prove that more aggressive pesticide controls will be needed.

Label information is of extreme importance. Pesticides are specific, meaning that they control specified insects but do not affect others. In addition, they are only appropriate for crops and situations listed such as ornamentals, tropical plants and greenhouses, fields, and interiors. There is no such thing as an all-purpose insecticide. Chemicals that may be used for outdoor gardens or greenhouses may not be labeled for, and therefore not safe for, interiorscape use. Similarly, pesticides labeled for interior plants are probably not suitable for home gardens. Never assume that a chemical may be used on edible crops.

Biological Controls

The usage of biological controls for insects has gained in recent years. Biological controls are natural enemies of plant pests, hence the name "beneficial insects." Since they are insects, their use may not be the best choice for all plant installations. For example, if a plant is infested with whiteflies, a predator whitefly can control them, but it necessitates the release of more flying insects. Once the host or bad insects are gone, beneficial insects lose their food sources and die. Biological controls can be effective for use in indoor atriums and conservatories. Many companies produce such insects for control of indoor ornamental plant insects. Some species perform better than others do depending on environment, whether in a greenhouse or atrium. Consult with the insect supplier about specific needs. Table 12-3 provides a reference list of indoor plant pests and their predators.

TABLE 12-3: Pests and Their Predators

Plant Pest	Predator Insect
Aphids	*Aphidoletes aphidimyza* Aphid Predator Midge
	Chrysoperla sp. Green Lacewings
Mealybugs	*Cryptolaemus* sp. Lady Beetle
	Lepidomastix dactylopii Mealybug Parasite
Spider mites	*Amblyseus* sp., *Phytoseiulus* sp., *Typhlodromus* sp. Predator Mites
Whiteflies	*Encarsia* sp. Predator Whiteflies

AVOIDANCE AND MANAGEMENT TECHNIQUES

The following considerations are provided in regard to avoiding insects before they arise and managing them after they are discovered. These guidelines should become a part of the maintenance technician's repertoire.

Avoidance Techniques

- Buy from reputable growers.
- Always quarantine new plants before they are introduced into an interiorscape.
- It is very important to scout for insects at every visit. If an infestation is caught in the early stages, the chances that it can be successfully controlled are greatly improved. Many infestations are introduced and peak in the springtime.
- Watch out for plants introduced into your interiorscapes without your knowledge.
- When plants become stressed, their weakened existence allows for easier, more rapid infestation. Plants, people, and pets can vector (carry) insects. While an aphid may not "bug" you, it is using you as a transportation service to its next time-share destination.
- Any houseplants kept outdoors for the summer should be treated with a few applications of insecticidal soap a few weeks before they are brought indoors for the winter.

Management Techniques

- Remove heavily infested plants swiftly and efficiently.
- Dispose of plants effectively. Do not give them to the client because they may forget to bring them home, aiding insects from the infested plant to spread to other plants in time.
- When disposing of infested plant parts or entire plants, carefully remove them and try not to shake the plant. Avoid letting the plant touch other plants around it. Contain it immediately in a sealed bag or take it directly to a dumpster. It is not unusual for a discarded, infested plant to reappear in its original site, put back in place by a well-meaning but uninformed person who intends on nursing it back to health.
- Oftentimes, it is better to discard a plant than to treat it. Consider the value of the plant first when making this decision, but do not take long to act. Insects spread quickly.
- Many interior insects do little harm to plants; they just bother people. Such nuisance pests can be detrimental to the bottom line of keeping plants indoors.

SUMMARY

The professional horticulturist's work centers on preventing problems before they arise. Taking the time and expense of meeting with growers, seeing their operations, and learning their philosophies helps interiorscapers to find important sources of quality indoor plants, free of insect pests. Upon receipt, inspect plants for signs of pests and quarantine plants for an extended period prior to installation in existing interiorscapes.

In established interior gardens, horticulture technicians must remember to scout for signs of insects at all times. If caught early, they are more easily controlled than if not noticed or never reported.

Small populations of insects are to be expected from time to time. Keeping them controlled through use of manual removal, water sprays, soap solutions, oil, or, if absolutely needed, insecticides are the ways of today's plant care technician.

Today's interiorscape is seen as a healthful balance of life. It is a slice of nature. The role of the technician is to keep balance between the beauty of nature and interior design.

CHAPTER 13

Diseases

INTRODUCTION

Plant diseases can be described as any pathogens and environmental conditions that adversely affect the normal growth and development of plants. Note that this definition involves not only pathogens that are living, known as **biotic** agents, but also **abiotic** agents that are environmental or cultural in origin and are nonliving. For example, abiotic agents that can impact plant health may be overwatering, placing a plant in range of ventilation drafts, or placing a plant in a vestibule to endure blasts of cold winter air.

Chemicals are available to control plant diseases, but keeping up with commercial compounds is a challenge. New chemicals are introduced but may not be labeled for interior usage. It is always best to check with a state pathology laboratory for assistance in identifying and controlling plant disease when maintaining interiorscapes of commercial importance.

The best approach to disease management is found in avoiding the conditions that foster disease. If plant professionals understand how to keep plant diseases out of the interiorscape, regularly dealing with them almost becomes a non-issue.

The Disease Triangle

Management of plant disease is easy until you find your plants are infected. Our approach as interiorscapers is avoidance of disease, so in order to avoid them, plant people should understand the classic theory of the disease triangle (Figure 13-1). In order for disease to grow, there must be a host. In the case of working in interior plantscaping, the host is a plant. This could be a practically microscopic plant in a tissue culture lab. It could be a seedling or cutting being produced in a greenhouse or a larger foliage plant in its final weeks of production in a nursery. It could also be a plant in a long-established interiorscape.

The second part of the disease triangle is the pathogen. We must assume that disease is always present, but not necessarily active or in a large colony. A puzzle piece or two are all many indoor plant diseases need in order to thrive. Unfortunately, it can be quite easy for plants to succumb to infection.

The right environmental conditions cause plant diseases to flourish. For the most part, it has to do with water, making moisture available to the pathogen to grow, but this is certainly not the only factor.

Disease Management Tips

When servicing plants, technicians should avoid splashing water on foliage (Figure 13-2). Water splash can vector disease as it is carried within traveling water droplets. Allow plant material surfaces to dry before pruning. Open wounds from pruning are entry ports for disease.

Do not crowd plants at any time, especially in planted installations. Many diseases flourish when air is still. Allowing air to flow between plants, even if spacing is just a few inches, helps to prevent diseases from spreading.

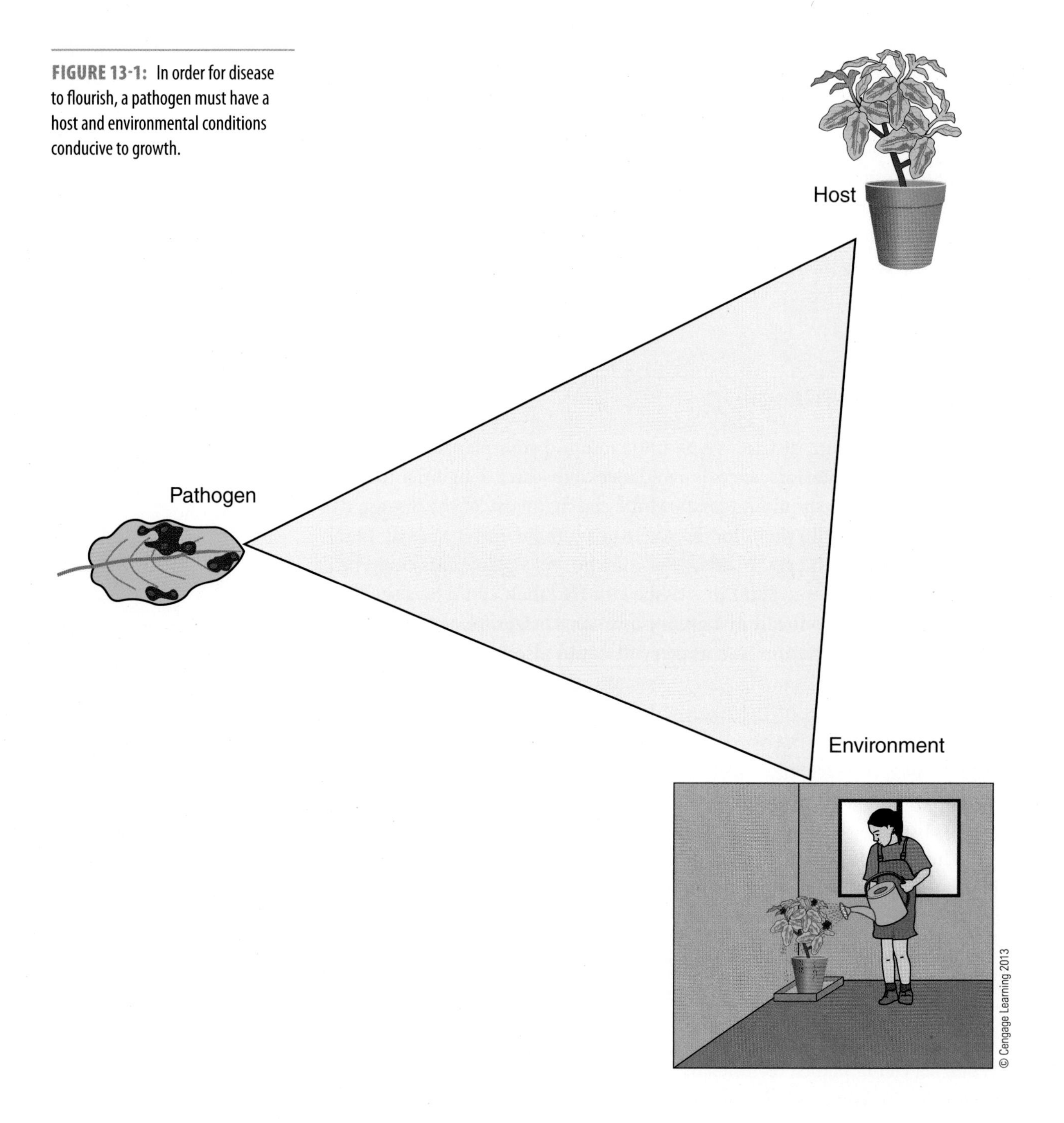

FIGURE 13-1: In order for disease to flourish, a pathogen must have a host and environmental conditions conducive to growth.

When a plant is repotted or otherwise transplanted, the height at which the plant is placed is important. For example, when offshoot bromeliads are potted, they should be placed a bit high in their new media. They are accustomed to well-drained soils and if placed too low, will be susceptible to rot. Most indoor plants are at their best when they are transplanted at the same height at which they were previously grown, level with the soil line. Keep the soil line about an inch or two below the rim of the grow pot to create a water reservoir. The water will percolate downward instead of running off the surface of the soil and off the rim of the container.

© iStockphoto/Olga Milkina

FIGURE 13-2: When watering plants, do not wet the foliage, but direct the water stream to the soil surface. Always keep leaves and flowers dry.

Sanitation is the key for keeping a groomed appearance and keeping plants healthy. Diseased flowers and leaves should be removed as soon as possible so that they do not come in contact with healthy tissue. Spent flowers should also be removed as soon as possible as they are prime hosts for disease. After removing diseased foliage or entire branches, immediately and carefully bag, box, or otherwise discard them so they do not come in contact with healthy materials (Figure 13-3).

Interiorscape technicians' hands, equipment, and clothing can vector disease. Techs should be instructed to wash their hands and clean their equipment frequently, especially after working with diseased plant material. Uniforms should be laundered regularly. Sanitize pruning shears or other tools with 1 part bleach to 10 parts water solution. Oil tools after sanitizing them to prevent rusting.

If a decision is made to use a fungicide, it will be more effective if used at the first signs of disease, before it has spread to other plants. Some plants may be harmed by fungicides. It is a good idea to test the fungicide on a part of the infected plant and wait one day to observe any deleterious effects.

When a plant falls prone to a disease, it is often advantageous to remove the plant altogether. Technicians should take special care to sanitize anything that came in contact with the plant such as jardinières, tools, and their own hands. If an initial plant and its replacements contract the same disease, use a different kind of plant for natural resistance.

© Stocksnapper/www.Shutterstock.com

FIGURE 13-3: Fully enclose diseased plant materials so that the disease is not mechanically spread to healthy plants.

Diseases Caused by Fungi

Living fungal diseases are not common in interior plants, but they do occur. Fungal diseases produce spores, which are vectored by contact or by air current. When the environmental conditions are right, fungal spores can penetrate leaf surfaces and flourish, causing deterioration of plant tissue. Common fungal diseases include anthracnose, root rot, crown rot, gray mold, and powdery mildew.

FIGURE 13-4: Symptoms of anthracnose, a fungal infection, are seen in this Red Maple leaf.

Anthracnose, *Collectotrichum, Glomerella*

Brown or tan spots appearing on leaves, petioles, and stems may signal an anthracnose infection (Figure 13-4). As spots mature, a concentric ring pattern can be seen. Anthracnose is easily spread by water splash. Plants should be watered at the soil level only, and watering in the morning is preferable because evaporation rates are higher than at night, even indoors. Control the disease by removing infected leaves and branches as soon as possible.

Root and Crown Rot, *Rhizoctonia, Pythium*

As noted in their common names, these diseases affect the underground portions of the plant first. Above ground level, symptoms such as leaf and stem wilt are exhibited, even after watering. This wilt is exacerbated by the practice of multiple waterings to counteract the problem. The additional water creates anaerobic conditions, limiting the necessary oxygen that roots need to survive, that help the diseases to thrive. Foliage yellows and dies, and then the entire plant ultimately dies. By the time symptoms are present, control may be difficult. This problem is better solved before it occurs. The employment of sterile media along with clean pots and other hard goods that come in contact with plants is important to stop spreading rots through generations of introduced plants. Overwatering is also a major culprit. Plants should not be allowed to stand in water unless the plant's native habitat is watery, such as in the case of *Cyperus alternifolius*, the Umbrella Plant.

Botrytis, Gray Mold, *Botrytis cinerea*

Gray mold earns its name from the appearance it takes on leaves and flowers. Grayish, fuzzy patches that distinguish *Botrytis* can develop on leaves and flowers (Figure 13-5). *Botrytis* does not flourish in environments where air can freely flow through plant materials, but it thrives in moist environments in still air. It also can thrive in cool temperatures.

This fungus is a problem in another sector of the floriculture industry in the phases between the grower and consumer levels, when flowers are packaged and stored. When moisture is trapped within packaged floriculture crops, *Botrytis* may grow and damage individual petals, flowers, or entire bunches of packaged flowers. Professionals and consumers should unwrap floral products as soon as possible to allow air to circulate around flower heads, foliage, and stems.

FIGURE 13-5: The vegetative mycelia of gray mold are seen growing on this plant material.

© iStockphoto/alohaspirit

Powdery Mildew, *Oidium* spp.

Powdery mildew appears as white, powdery spots on leaf surfaces (Figure 13-6). One of the best ways to prevent it from spreading is to remove the infected leaves and place them in a plastic bag. Take care not to touch healthy leaves or other plant parts once you have touched an infected leaf. Though not harmful to humans, your hands can spread the disease to otherwise healthy plants you service. Since fungal diseases are impossible to see, it is important that plant service technicians get into the habit of washing their hands regularly.

FIGURE 13-6: This Lilac shrub leaf is heavily coated with powdery mildew. Interior plants having this infection are more likely to have smaller patches of the fungus.

© Cengage Learning 2013

Fungal Leaf Spots

There are several genera of fungi that thrive on indoor plants. Sometimes, infections cause oddly placed spots on leaves and potentially other parts of plants. *Fusarium* causes rotting of leaves and roots, which turn mushy and brown. *Erwinia* leaf rot may flourish under warm, moist conditions. *Fusarium* stem and crown rot can affect bromeliads and many other plants. *Xanthamonas* can spread rapidly through the vascular system of plants like *Aglaonema*, *Anthurium*, and *Syngonium*, which are more susceptible. These diseases are a greater threat

GREEN TIP

Recycle or reuse black plastic grow pots. They should be sanitized prior to reuse to avoid disease or insect vectoring by scrubbing with a soapy water detergent and allowed to thoroughly air dry.

at the grower level where hundreds of plants can be affected at one time. Indoors, it is best to remove infected leaves and avoid vectoring. If plants lose a good appearance, remove and discard them away from healthy plant materials.

Diseases Caused by Bacteria

Bacterial diseases thrive when surface moisture is present. A pathway for this type of infection is made when plant material is damaged or stressed. Leaves may be torn or stems may be lacerated during installation or maintenance, or by malevolence. These wounds become entry ports for bacterial infection. Already sick plants are also prone. Bacterial leaf spots and bacterial blights are numerous and proliferate in humid, warm environments. Think about spaces that are warm and humid. Indoor pools, spas, and workout facilities offer better conditions for bacterial infections than do offices or retail locations. Technicians maintaining plants in warm, humid sites should always avoid wetting foliage and should keep all necrotic leaf margins trimmed and clean because the dead tissue is optimal for bacteria. If plants must be pruned, conduct the work at times of the day or week when humidity levels are lower, or move the plants to a space with similar light intensity and duration until the wounds are healed.

Soluble Salt Toxicity

Abiotic soluble salts may accumulate in soils and over time may become toxic to plant roots. High levels of salts may actually attract water away from roots causing them to die. Stressed and dead roots cannot take up water, so foliage will also suffer by becoming yellow and wilting. Overall, the plant will lose a healthy appearance and require replacement.

Since these symptoms are specific not only to soluble salt damage but also to numerous other problems, a soil test by a reputable laboratory is best. It is a good practice to conduct soil testing annually for installations that use valuable plants, are on the property of high-profile clients, or have high-dollar maintenance contracts.

Some plants show sensitivity to low levels of dissolved solids (soluble salts) by exhibiting tip burn. Necrotic leaf tips, especially in narrow-leafed plants such as *Dracaena deremensis* 'Warneckii' or *Chlorophytum comosum,* Spider Plant, are more prone to this type of injury. The way to solve this problem is to carefully trim away the necrotic leaf tissue, taking care to follow the natural pattern of the leaf. Over time, additional necrotic tissue may appear which should eventually be trimmed until the point when the entire plant should be replaced. The decorative life span of an indoor plant is usually measured in years rather than weeks. Such is the case of tip burn due to soluble salt damage.

Looking more closely, the appearance of a white "salt" may appear near the soil line on the inner surface of the pot. If the plant is in a clay pot, the salt leaches through and collects on the exterior walls of the terra cotta. Salts can appear on the edges of the pot's drainage holes.

Sometimes it is more practical to leach salts from potting media rather than to replace media. Multiple waterings during a short time span, perhaps one day, allow salts to stream through the media and wash out through the pot's drainage holes. Excess leachate should be drained away as soon as possible from the plant. This type of preventive work is messy and therefore cannot easily be performed at the interiorscape site. It may be beneficial to temporarily replace the plant to be leached with another plant. As part of regular maintenance, plants may be rotated this way, some on display and some at the interiorscape home site.

Another way of countering soluble salt problems is to completely repot the plant in a freshly scrubbed and rinsed container with new, sterile potting media. Sometimes, the work is worth the effort in maintaining plant health and customer satisfaction.

Common Symptoms of Plant Disease

Sometimes, identification is a challenge due to the way symptoms are described. When trying to match observed problems to possible diseases, the following symptoms may be listed.

Blight—quickly developing collapse and death of part of or entire plant
Canker—concave or convex dead tissue areas on stem
Chlorosis—yellowing, mostly in leaves, because of chlorophyll cell death
Dieback—dead shoots
Distortion—leaves and shoots growing in shapes that are not characteristic of plant
Leaf spot—mostly appearing on leaves; distinct, dead, or dying patches of tissue
Mosaic—like a tile mosaic, small areas of dead or dying tissue among healthy leaf tissue
Necrosis—plant tissue death
Rot—soft, watery dead and dying areas appearing in leaves, roots, and shoots
Stunting—shoots or entire plant not growing to full potential
Wilt—lack of turgor pressure in plant cells causes leaves and stems to appear floppy

SUMMARY

Preventing disease predicaments is the safest and best way to control the problem. By the time symptoms are noticeable, it is usually too late to easily stop disease from killing the host plant. Keying plant diseases is an art because of symptom similarities. Spraying with fungicides can be futile if a disease is misdiagnosed because effective treatments are often specific. Today, we are using very little chemical control, counteracting plant problems with forethought, scouting, sanitation, and replacement.

Most diseases love moisture, still air, and low light. Always avoid stressing plants with too much water. Space plants so that they enjoy air circulation with no water splash on foliage. If planting directly into beds, level root balls at the same height at which they were originally grown.

When technicians and other plant lovers avoid the third component of the triangle, plant diseases will rarely be a problem. Choosing hardy and healthy plants from reputable growers who consistently use appropriate cultural practices fortifies the plantscaper's work.

SECTION FOUR

BUSINESS

In the world of business, nothing can progress until there has been an exchange of money for products and services. This may seem simple, but the hierarchical intricacies of business are complicated, all tied together by fulfilling consumer needs. When money is exchanged for plant placement and care, expectations arise on the part of the consumer. Interior plantscaping is a business dependent upon the flow of products and services between growers to wholesalers, to retailers, and, ultimately, to consumers. The following chapters shed light on the structure of an interiorscape business and the work accomplished by its people.

CHAPTER 14

Business Management

INTRODUCTION

We separate the love of interior plant culture from the profession of interior plant culture. The word *amateur* holds the Latin root word *amat*, which means love. Amateurs practice their craft but do not create products or offer services for income. People in this profession not only have an affinity for indoor plants but also earn income through the design, installation, and maintenance of plants and related products and services. The structure of a business depends upon finding customers, providing quality plants and care, and maintaining their satisfaction at a fair market value. In order to accomplish these tasks, different company structures with various job positions seek to fulfill objectives.

Many people consider careers in horticulture and find reward through working within a plant-to-people environment. Keeping plants indoors remains a mystery to many consumers, yet they desire natural beauty to be a part of their work and domestic environments. For the most part, these professionals do not have the time or the interest to care for plants. They are much too busy focusing their attention on other goals, but they do have the financial means to hire pros who know indoor horticulture. Interiorscaping can be a lucrative profession if you gain knowledge and business acumen to succeed. Working as a horticulture specialist is fun. Making a sale and pleasing customers is exciting when combined with the satisfaction of providing beautiful plants and containers to enhance people's surroundings and, in a special way, improve their lives.

The Business of Interiorscaping

No one can be successful in professional interiorscaping merely because they love plants. Although seeing healthy plants improve an interior setting is core to the business and probably the primary reason someone gets involved in professional interiorscaping, being an interiorscaper means much more. Many hours of networking and traveling about a service area are necessary to start a company and keep it running. Beyond these efforts, a company owner must possess get-up-and-go. Successful owners have self-initiative. Instead of spending a day lounging, they prefer a day of activity, not only thinking about projects, but also putting them into practice. A positive, friendly attitude is completely necessary to own or manage a company over the long haul. Customers know when they are valued, and this feeling must be maintained to not only attract clientele but also to keep them.

In order to be successful over time, interiorscape firms must focus on excellence, with both quality plants and superior service. All plants sold and maintained must always appear healthy and attractive. Good plant displays generate word-of-mouth advertising. The more people who see terrific plant installations, the more potential customers you will attract. Along with great products, interiorscapers should strive to offer excellent plant care. Services should be made on time and in a skillful manner. Staff members must maintain a professional attitude. These attributes build your brand. Although on the surface a brand may seem to be a company logo, the logo symbolizes what your company is perceived to be.

Job Positions

An interiorscape company contains five basic departments, which include sales, design, installation, plant care technicians, and bookkeeping. In the case of a new, small business, one person may operate these departments. Larger firms may have dozens of employees in a single department.

Sales

Sales staff members are some of the first personal contacts between the interiorscape company and the client (Figure 14-1). Successful sales personnel are good listeners, making a connection between client needs and company needs. Forever mindful of generating sales, they keep in mind what it takes to install and maintain plants that will consistently please customers. They are professional in appearance and action and can answer questions about plant installation and maintenance quickly and accurately. They provide price quotes to clients in a timely fashion and use approved contracts for clear communication among all parties.

© Stephen Coburn/www.Shutterstock.com

FIGURE 14-1: Sales staff travels to meet potential interiorscape clients. They are often the only face of the interiorscape service provider.

Design

Interiorscape designers have the ability to translate the right geometric forms into appropriate plants for displays in approved spaces. Designers may use various methods of expressing design plans including drawing with pen and ink, markers, watercolors, or illustrations using design software. Designers notice architecture, interior design styles, and elements along with taking cues from the client. They seek to provide an interiorscape that the client envisions within a feasible budget.

It is common for interiorscape firms to combine design and sales staff together. The person who sells an installation is also the designer (Figure 14-2). Some people are better at design than sales, others are better at sales than design. The important thing is to improve upon both tasks, especially the weaker of the two.

© StockLite/www.Shutterstock.com

FIGURE 14-2: In most instances, interiorscapers both sell interior plant installations and care services and create the design layout for their clients.

Installation

Installers receive, organize, load, transport, unload, and position plants and containers, and they clean up when the job is completed. Maintenance technicians sometime help with installations. Interiorscape departments within horticultural companies have a greater pool of crewmembers to help (Figure 14-3). Larger jobs may involve the help of independent contractors to supplement full- and part-time staff. It is important to remember that plantscape installations are limited in comparison to maintenance visits, which are frequent. An interiorscape firm cannot operate solely from installation income. The revenue generated from maintenance contracts drives the company income over time. Crewmembers must be patient and keep attitudes professional in demeanor. They will be working with management, designers, sales, and staff. Client representatives may be present when installations are occurring, so the crew should always remember to treat everyone with respect.

FIGURE 14-3: Workers install trays of plants for a green roof.

Plant Care Technicians

Technicians (techs) provide the care to keep installed plants looking their best at all times (Figure 14-4). They should not only provide routine services such as watering, dusting, turning, and trimming but should also regularly evaluate plants' aesthetics (see box on Plant Care Technician's Maintenance Tasks). They must make judgments and report their findings to management, most often via company forms. Most interior settings provide very low light levels that can sustain plants over a period of months, perhaps years, but do not provide intensities for shoot development. Indoor plants should be thought of as living sculptures that improve productivity, air quality, and ambience of the interior environment. Plants should be maintained as contributors, not detractors, to the company or private

GREEN TIP

Maintain proper inflation in tires used on installation crew and maintenance crew vehicles. This improves fuel economy and saves money.

FIGURE 14-4: The plant care technician is the link between the interiorscape company and the client, providing products and the best service possible.

GREEN TIP

Interiorscape personnel rely upon transportation to service accounts. Firms should consider use of flex-fuel vehicles, biodiesel, or electric cars.

residence. Wilted plants, dead or yellowing leaves, or any other problems that foster negative publicity must be avoided at all times. Poor-quality plants are bad for the client's image, the interiorscaper's bottom line, and the interior plant industry.

PLANT CARE TECHNICIAN'S MAINTENANCE TASKS

Applying fertilizers if needed
Dusting and periodic washing of leaves
Dusting or cleaning plant containers
Keeping accurate water levels in sub-irrigation systems
Removing trash left in pots or on mulch surfaces
Replacing unhealthy-looking plants with new plants of similar size and value
Scouting and reporting signs of insect infestations and diseases
Trimming necrotic leaf margins
Watering individual plants
Replacing or finessing mulches or coverings

Bookkeeping

A part-time or full-time staff member may be hired to keep track of the interiorscape company bookkeeping. This would include accounts payable, such as to wholesale greenhouses and container and supply companies; accounts receivable, the income from maintenance accounts and installations; purchasing; and payroll. Much trust is placed in accounting employees to keep accurate books and perform their work so that cash flow is expedited (Figure 14-5).

FIGURE 14-5: Mid-size and larger horticultural companies are able to hire part-time and full-time bookkeeping staff.

Independent Contractors

From time to time, it is necessary to secure extra help when specialized labor or extra hands are needed. One example may be the need to hire additional workers for a major installation of plants in a public building occurring on a Saturday night or hiring floral designers to create a holiday display in a mall (Figure 14-6). When hiring independent contractors, employers do not have to perform the following: offer fringe benefits, pay unemployment taxes, pay half of FICA (Federal Insurance Contributions

Act tax, a United States payroll tax imposed by the federal government on employees and employers to fund Social Security and Medicare—federal programs that provide benefits for retirees, the disabled, and children of deceased workers), pay other state and local taxes, or withhold taxes from their pay.

There is definitely a financial savings with "spot labor," but they can be difficult to find because they are few in number and are in demand. Employers must complete 1099s (tax identification form for miscellaneous income) for them, pay a commission or pay on a per-job basis, request a bill from them stating the agreed-upon amount of pay, and provide them with a contract stipulating job objectives, dates, and time of employment. Remind independent contractors that they get travel and entertainment deductions on their income tax returns, so they should save and turn in such receipts to their accountants.

Any employer contemplating the hire of an independent contractor is wise to contact their counsel and accountant prior to the appointment, as soon as it seems the need will arise. It is also prudent to draw up and sign an agreement that specifies the terms of the independent contractor relationship. This may include the duration of employment, rate, and frequency of pay, but is not limited to only this information. Remember that contracts are forms of communication that help all parties in understanding objectives.

FIGURE 14-6: An independent contractor can provide products and services not always offered by interiorscapers but necessary to fulfill contract specifications.

Professional Attitude

It is fun and rewarding to design an outstanding interiorscape, one that is creative and edgy, something that is truly different. It is even more rewarding to sell the thing. Satisfaction of a job well done occurs when a monetary exchange is made; a fair value is traded for an outstanding product. This seems simple, but in actuality, it is quite difficult because personal interaction is part of the mix. Professional attitude paves the way for positive, successful, interiorscaper-to-client interaction.

Think about operating an interiorscape business from the standpoint of an owner. As an employer, what are the characteristics of an ideal employee? Specific characteristics might include someone who consistently gets to work

Some of the words used to describe successful people who possess the right attitude include:

- Creates win/win outcomes
- Disciplined
- Fun
- Helpful
- Knowledgeable
- Motivated
- Skillful
- Solution-oriented
- Sound
- Successful
- Truthful
- Unsinkable
- Upbeat

on time, or better yet, a few minutes early. If a client would like plant services to conclude before it opens to the public at 8:00 a.m., the technician should be finished with all work and the area cleaned by 7:50 a.m. Better employees dress appropriately for work with proper footwear, not flip-flops or unsafe, open-toed shoes. They take their caps off indoors out of respect for other people. They bathe regularly. They do not take drugs or arrive at work under the influence of alcohol.

A professional attitude is a successful attitude, not one bogged down by feelings of failure. Successful people see opportunity in everything and often say "yes" more than "no." A negative attitude might make us say things like "I don't want to travel to that account, they are too far away," or "I don't do big cleaning jobs." In actuality, plant maintenance is one of the chief reasons why we have interiorscape careers in the first place. Much consumer value is placed on keeping plants looking great. Imagine the way plants give life to the interior.

Give thought to a company uniform. All entities of the interiorscape company should be identifiable at first sight due to their distinctive uniform. Plant-care technicians should be provided with a minimum of several shirts of the same design and color combination that may be paired with neat, tailored khaki slacks or clean jeans with no tears or holes. Of course, variations in weather and seasons result in appropriate dress-code variations, but overall, a uniform look in the company's colors aid in brand recognition. Torn slacks, faded shirts, or a complete lack of a uniform have a negative effect on brand building and could signal the need for company organization. A client might ask, "If you do not care about your appearance, how are you going to care about plants?"

Now, think like an employee within an interiorscape company. Perhaps you are a new employee with a small amount of experience or none at all. One of the best things you can do is to make yourself indispensable to the company. Think ahead when situations arise to form successful solutions. When a simple plant replacement must be made and closing time is in 45 minutes, deliver and install it rather than leave it for someone else to do. What would it take for your boss to think you are irreplaceable?

A professional attitude says, "I can do this." One of America's greatest playwrights, Tennessee Williams, used to tell himself that writing was easy even though it is not. Can you imagine writing so prolifically, let alone with such emotion as Williams? It is not easy, but if we tell ourselves it is, we will at least pave the way for an attempt, which is a mile ahead of never starting at all. At all costs, avoid "I can't," "I won't," and "I don't want to." Such phrases are debilitating over time. We do not sell plants so much as we sell living interior enhancements that provide an emotional response. It takes someone special to be aware of the important balance between plant health and environmental beauty (Figure 14-7). A certain set of characteristics form a quality horticultural professional. They consistently work toward win/win situations when challenged.

FIGURE 14-7: There is much satisfaction in working hard to achieve a goal.

Starting an Interiorscape Business

People have many reasons for wanting their own business. They may wish to develop income doing something they love, thus experiencing more fulfilling lives. They may have a desire to make their own decisions based upon what they feel would be the best in products and services. Sometimes people feel that business ownership equates higher income, but this is not necessarily the case.

Get Experience First

If you think a career in indoor horticulture is for you, it is best to work for a few different firms first. Get your feet wet and learn the ropes. Students may perform occupational internships or cooperative learning experiences as part of degree programs. This is a terrific time to gain professional experience. Scholarships and grants are available to aid in funding expenses incurred during an internship. Often, placement in horticultural enterprises is available. If possible, student interns should consider internships where the most valuable learning experience will occur, even if this means traveling to another state, region, or country. Your learning will be put to work if you see trends that are far ahead of your current market area.

People who are out of school and are interested in learning more about interiorscaping should seek jobs directly with interiorscapers or similar enterprises and departments. Entry-level positions offer lower pay, but the opportunity to see how a company works from the inside is valuable. All experience will show that this work is not easy and long hours are required. Some days will start very early and end very late. Some work will take place during the night hours when traffic is slow, for example, in the lobby of an upscale hotel or mall.

Many entrepreneurs have found starting up an interiorscape firm really does not require much capital. It can be operated as a home-based business. This has a tax benefit because part of the residence will be devoted to office space, supply storage, and perhaps square footage for holding plants. Starting a business in your home requires less up-front capital and has a lower overall cost. It would be cheaper to store hard goods in your basement or garage, or even rent a space at a storage facility. Before a home business is established, check with local government offices regarding home business ordinances and regulations. Many local municipalities do not allow signage and have concerns about parking and extra traffic.

Conduct a SWOT Analysis

When considering any business venture, an enlightening way to organize your thoughts can be found in developing a SWOT analysis. The acronym SWOT stands for strengths, weaknesses, opportunities, and threats to the business success. For the most part, strengths and weaknesses are internal to the business. Strengths may include plant-care experience, supply source contacts, and ownership of a reliable vehicle. These reflect on what the entrepreneur will bring to the business. Weaknesses are also internal and could include lack of business capital, time limitations like being the sole caregiver to an elderly parent, or being shy when it comes to drumming up business. It is important to remember that millions of people have similar weaknesses as you and they have successfully overcome obstacles to achieve business success.

Opportunities and threats are external to the business. Opportunities must be explored, such as special loan programs for minority business owners or requests for your products and services before you even open. Threats to growing an interiorscape business are both seen and unforeseen. An unpredictable change in the overall economy may force some accounts to limit the number of plants and associated maintenance fees. Competitors may try to undercut established maintenance fees resulting in your performing the same amount of work for less money.

Develop a Mission Statement

Every business needs a brief mission statement that delivers the purpose, objectives, and principles that drive the company. When taking time to develop the mission statement, think in terms of far-reaching ideals rather than just what you want to accomplish this year. For example,

> Our objective is to supply and maintain healthy, durable plants for commercial and private interiors in the greater metropolitan area. Through quality plants, products, and services, we desire to become our area's top interiorscape service.

Apply for a Business License

Check the local and state government web sites to obtain a business license. There is a fee for the license, which varies from state to state. This license registers the company for the collection of tax dollars.

Market Your Business

It is important to have a web site for your business as it is often the first time a potential client comes in contact with your company. At the very least, it is an interface for basic information including your company name, physical and mailing address, email address, phone numbers, and product and service information. Because interiorscaping is a design-based business, photographic images are extremely important. A web site can be a portfolio of your installation work, and after each job is completed, images depicting the job in the most attractive way should be posted to the web site. You should also consider testimonials from customers.

Pricing lists are not necessary, however, and should be avoided. This is due to cost fluctuations, pricing breaks for larger jobs, and the need to adjust contract bottom lines due to emerging competition. This is something that you will want to handle face to face, after seeing the client's space and judging exactly what it will take to give them the best installation and maintenance contract for its money. Link your email to the web site in order to receive direct inquiries. Try to follow up on business inquiries as soon as possible, preferably within one business day or less.

A basic web site is better than none at all, but it is best to have a competent, talented web developer design your site. Some companies try to conserve precious capital, designing web sites themselves with free templates or having an amateur take on the task. These products do not shed positive light on your company and make your products appear to be cheap and your design work ineffective. As is true with many things, the least expensive means are not always the best.

It is not helpful to the consumer (indeed it can be aggravating) when a web site contains outdated information. Imagine featuring the front page of your site with an image of a Christmascape loaded with red poinsettias posted in October but remaining in March. It makes your company appear out of touch. Be realistic about the frequency of posting new images. Can you make time to update your site yearly, semi-annually, or monthly? It takes time to post images and other information, even if your web site is professionally maintained.

It is better to avoid fancy effects and music that slow down the rate of site loading. Keep in mind your web site's audience. These professional people want beautiful, healthy plants that will improve their corporate appearance.

Use local networks to market your business. Join and speak at civic groups to let people know who you are and what you do. Civic groups such as Kiwanis, garden clubs, and business associations frequently seek individuals to speak about a topic. Consider a short program on plants for clean air or the top-five hardest-to-kill indoor plants. Bring examples of healthy, desirable plants with you along with handouts that highlight your talk, and, most importantly, bring plenty of business cards that list your name, phone number, email address, and web site address.

By delivering such a talk, you deliver factual, useful information to listeners and send a positive marketing message that you are an expert in interiorscaping as well as a local professional who cares about civic good. These talks may win you new installations or maintenance contracts with commercial spaces as well as private residences. Do not underestimate any group, even if you feel it is insignificant. A garden club may not have a CEO as a member, but it may hold members with the means to have their interior plants installed and maintained by a professional or members with spouses who are the decision-makers when it comes to choosing an interiorscape firm.

Make efforts to meet with landscape architects and owners or managers of local nurseries and garden centers. Think about the waiting rooms of your own doctors, dentists, lawyers, and other entities that need plants. Restaurants and hotels, large, small, or in between, need plants to greet guests and to keep a fresh look. It is worth the time and cost to develop a paper flier that lists the benefits of your company's plants and services. This silent salesperson may be left with management as a reminder. It is a more detailed calling card than a plain business card, and it is necessary in order to provide visual images beyond your well-chosen words to develop sales. Web sites are terrific and necessary when people are seeking information online, but personal introductions make important connections in clientele building.

What about those companies that already have interiorscaped spaces? You may request to submit a maintenance quote just before their current plant care contract expires. Any space where plants can be or are maintained could potentially be your space if you are assertive about seeking and securing new business.

Professional Organizations

Through interaction with professional organizations, you can meet hundreds of professional interiorscapers from throughout the country and the world. Organizations such as ASLA (American Society of Landscape Architects) and PLANET (The Professional Landcare Network) offer the chance to network with seasoned professionals who can help you get in contact with better suppliers and offer advice on all types of business questions. Many of the people you meet have experienced similar challenges and could provide beneficial solutions.

If interiorscaping is your livelihood, it should be on your mind all the time. Do not be afraid to make conversation about interior plants and their importance wherever you trade, from clothing stores to doctors' offices to salons. When appropriate, give a live plant in a terrific container as a gift to business associates, friends, and family. Do not hesitate to give a desk-sized plant to potential clients. One plant may sell hundreds!

The most challenging part of an interiorscape firm is working with people. When poling employers and staff regarding continuing educational needs in horticulture, one of the topics that consistently rises is "how to work with difficult people." Indeed, this is one of the major tasks in any setting where people must work together to solve problems. Clear communication of objectives takes time but produces beneficial outcomes.

Types of Small Businesses

There are many types of ownership structures in the business of horticulture. This section introduces the basics of company ownership, starting from the simplest and working toward models that are more complex.

Sole Proprietorship

A sole proprietorship is common in horticulture, where the business and the individual are one and the same. It is simple and relatively inexpensive to hire a certified public accountant (CPA) to oversee your business and personal bookkeeping. He or she can help you develop a system to keep track of income and expenses. Income and losses are reported on the sole proprietor's income tax report. The same individual is responsible for court judgments, including being sued and debts. Sole proprietors pay taxes on their business income along with any other derived income. Partnerships are similar to sole proprietorships but include ownership by two or more people. Business partners must mutually respect each other's opinions and decisions in order to flourish. Sometimes, business partnerships build stronger friendships while some friendships are ruined by partnerships.

Limited Liability Company

Limited liability companies (LLCs) have grown in number in recent years. An important difference in an LLC over a partnership or a corporation is that LLC members are limited on the liability of the members for debt payment and court judgments. Their personal property cannot be seized in order to pay a penalty. This option for interiorscapers may offer its greatest benefit if a company is sued by a client. Personal assets of LLC members cannot be taken away. Note that the people involved in an LLC are not partners or shareholders, but members. They are generally expected to pay employment taxes from their LLC earnings. When a member of an LLC leaves or dies, the LLC is dissolved and a new entity must be organized in order to conduct business.

To explain further, say an employee of a tenant in a large office building your company services slips on a water spill made by your interiorscape technician, falls, and becomes severely injured. If you are not fully insured for such an accident and you are a sole proprietorship, your assets such as your savings, your personal vehicles, or your home may be at stake in order to pay costs and penalties. The same is true for a partnership, except more than one person is held liable. In an LLC, such personal assets are protected, but associated operating costs are higher due to legal and financial complexities.

Corporation

Corporations provide a different kind of structure for bigger businesses. There are two types of corporations available, the S-Corporation and the C-Corporation. S-Corporations are an independent legal and tax entity and are separate from the people who own it, called shareholders. This separation means the corporation's income or losses are divided among and passed to its shareholders. Applications to develop corporations are filed with the state and developers must request S-Corporation status or the body automatically becomes a C-Corporation. Shareholders must report the income or loss on their own individual income tax returns. This concept is called single taxation.

In short, S-Corporation employees are taxed on their salary income, but not on corporate dividends. Shareholders are not liable for debts. If the corporation is taxed as a C-Corporation, it will face double taxation, meaning both the corporation's profits and the shareholders' dividends will be taxed. In other words, the income of a C-Corporation is taxed at the corporate level and dividends paid to shareholders are also taxed as income.

Most interiorscape firms are small businesses that are operated as sole proprietorships or partnerships when two or more persons go into business together. As a company grows in size and stature, developing an LLC may be a good option to protect the owners' assets in case the company is sued.

Financing a New Business

If possible, it is best to support a fledgling business with capital in-hand, in other words using money from your savings to finance the purchase of plants, supplies, and daily expenses including fuel. It may take more money than you have on hand, in which case a business loan may be taken. The U.S. Small

Business Administration web site offers information about possible loans and requires applicants to submit financial statements including:

- balance sheets from the last three fiscal year-ends
- income statements revealing your business profits or losses for the last three years
- cash flow projections indicating how much cash you expect to generate to repay the loan
- accounts receivable and "payable aging," breaking your receivables and payables into 30-, 60-, 90-, and past 90-day-old categories
- personal financial statements from you and your business partners listing all personal assets, liabilities, and monthly payments, as well as your personal tax returns for the past three years

Bankers want this information as well if you are seeking a loan from a local lending agency, and they may also require a resume detailing your professional and educational background. Some lenders ask for a brief verbal presentation with visual aids about the company, its location, products, services, and goals. Of course, the first step in securing a loan is to consult with your accountant along with possible lending agencies in order to develop the necessary document portfolio needed for loan consideration.

Bank Account

In starting up a small business, it is necessary to open a business checking account separate from personal accounts. This business account should be opened with what is termed a "fictitious name"; in other words, the name of the business rather than the business owner. The fictitious name is often listed after the name of the owner with the initials "d.b.a." (doing business as). Separate business and personal accounts are important in keeping accurate records for the owner and professional accountant to chart company progress.

Bookkeeping

A number of easy-to-use small business software accounting programs are available. Check with your accountant for suggestions on software programs so that you are working with his or her preferences as well as your own. It is a good idea for the owner of the company to take care of bookkeeping or to have at-a-moment access to it in order to know your company's growth or shortfalls. Many people get involved with interiorscaping because they love plants, but it is also necessary to love, or learn to love, your company's books. The two go hand in hand in order to attain financial success in the industry.

Taxes

It is necessary for the company to file income tax returns, with payments made at assigned deadlines throughout the year. The sole proprietor or partners must also pay self-employment taxes on income earned from the business. Independent contractors must file their own income tax returns with yearly income information provided by the interiorscape company. It is highly advisable and necessary to work with a CPA. CPAs are qualified to work with small businesses and aid greatly in the calculation of taxable income. The CPA may also act as a consultant with the business owner in providing advice on how to organize bookkeeping for ease and accuracy. They are required to keep abreast of all changes in tax laws, which would be nearly impossible for a busy, small business owner. Ask the accountant to provide you with an estimate of his or her tax preparation services in order to budget for the amount. The cost of fees associated with a CPA's services is well worth the investment in keeping your company organized and within legal guidelines.

Pricing

It is difficult and potentially misleading to provide exacting information on pricing in this textbook, even though many readers would value this information. One of the pitfalls of setting up pricing guidelines or formulas is that people would automatically follow them as an industry standard. Pricing plants for retail along with additional products and service levels is tricky, requiring frequent, subtle adjustments. If everyone was given a rule book for pricing, your competition could easily undercut your prices because they know your rules.

In general, when figuring retail prices, the old adage "whatever the market will bear" holds. You should charge the *highest amount possible* for products and services *with knowledge* of what your competition is charging. While this answer is best served by working in the industry first, it is possible to find some information by inquiring about installation and maintenance costs from companies with established interiorscapes. A few doors may be shut in your face, but a professional demeanor coupled with sincere interest in building the best interiorscape company in the area will go a long way with clients.

Another good prospect in learning more specifics about pricing and mark-up of horticultural goods and services is through plant and hard good suppliers. Wholesale growers have close relationships with many of their clients and share information about retail price ranges. While sharing names and specific information can be deemed unethical, sharing "ballpark" figures, price *ranges*, and generalized information is deemed helpful to both the retail buyer and the wholesale supplier. The wholesaler may view this as a way of teaching and cultivating a long-term customer. In turn, you as a buyer are forming the good foundation of securing a reliable supplier. Securing reliable suppliers who are willing to help with requests as much as possible is an overlooked part of the sales/installation process. Many interiorscapers forget the time it takes to find the right plants for the job, also forgetting that time is equivalent to money.

It is easy to recognize that we are selling and maintaining plants as interiorscapers, but we have to remember the human factors as well. Company employees are our greatest assets. We spend a great deal of resources locating, training, and maintaining staff members in the interiorscape industry. Employees are needed for sales, installation, and all-important technical maintenance, and their work deserves reward. The student or new hort technician does not always realize how much an employee truly costs. Beyond an hourly salary, employers must also pay Social Security taxes, unemployment insurance, disability insurance, and workers' compensation insurance, and they may contribute to other benefits such as health insurance and occasional education expense for manuals or short courses.

Precious funds spent on education and certifications are important because they help a company make money over time. Building professional contacts becomes much easier through participation in conferences and trade shows. Interiorscapers are more willing to share "business secrets" when they know your company is located so far away as to not be competitive with them. This is a great way to learn about pricing for installation and plant replacement. In turn, you may have information that is valuable to them, thus creating a win/win relationship. After making such friendships, advice is just a phone call or email away.

Many interiorscapers double the wholesale price of plants and add a bit more to cover the cost of shipping. These shipping charges are highly variable due to distance from production areas to retailers and the fluctuating cost of fuel. This 2:1 markup may be adjusted downward if a job is large, with many plants installed, or it may be adjusted upward if there are just one or two plants installed. Additional fees are often charged to cover the cost of installation labor. Some installations are made during business hours, perhaps after furniture is moved into a newly constructed space and deadlines are ahead of schedule. Luxury of time is the exception because larger installations are often made after hours, on Sundays, or during the dead of night to avoid slowdowns associated with automotive and pedestrian traffic. It is much easier to install and maintain plants without people about, but labor prices may escalate at off-hours.

In the development of a maintenance contract, features other than existing plants and hourly pay rates should be factored. When surveying an interiorscape to make a maintenance program bid, scout the space and look for water sources. Are they near the plant locations or are they remote? Is the available light conducive for growth or maintenance? If lighting will encourage growth, it may take longer to maintain the site per month or year due to pruning or soil refreshing. If the site is in an urban area, it may be necessary to pay for parking.

Insurance

When devising the retail pricing of a single plant or a large installation, it is easy to "give away" the item or project. For that matter, most horticulturists are down-to-earth people who just happen to love plants more than money. This may be the case, but it is not true of the rest of the world, so you must protect yourself. Necessary insurance costs money, and there are several ways of protecting yourself and your assets. Automobiles and trucks must be insured. General liability insurance to aid employees or clients injured due to your company's negligence and disability insurance if employees cannot work for a period of time are also necessary expenses.

Many service providers advertise the fact that they are licensed, bonded, and insured. Licensing means they hold permits with their local government office and are competent to perform specific types of work, for instance plumbing contractors who can successfully install sinks, toilets, and associated piping. Local municipalities generally do not license interiorscapers, but individual employees can hold state pesticide application licensing and industry organization certification. Bonding and insurance play roles in company security. When a company advertises it is bonded, it means the company has money set aside in control of the state that would be paid to clients in the event a claim was filed against the interiorscape firm by a client. For example, if a court determines your plant technician stole computers from your client, the bonding company would pay the client for the loss. When a horticulture firm advertises that it is insured, it means that it will cover medical expenses if one of its employees is injured while on a client's property. If an unfortunate installation crewmember got hurt installing large-scale trees in a mall, your insurance, not your client's, would pay associated medical costs. All interiorscape companies should be insured, and for companies with multiple plant technicians, bonding provides the contractor and the client with peace of mind.

Storefront

© Gordana Sermek /www.Shutterstock.com

FIGURE 14-8: Keep customers returning by having a constant supply of attractive products available for sale.

See your business through the eyes of others. If you own or manage a storefront company, such as a nursery or garden center, resist the urge to constantly use the employee entrance. Inconvenience yourself for a few minutes and use the main entrance of the store. What do customers see as they enter? Are healthy, attractive plants on display? Are they well maintained or are they water-stressed with yellow leaves? Are stylish containers in a variety of sizes and colors part of the product mix? Remember, you cannot make a sale from an empty cart. There must be a constant supply of attractive products for sale in order to keep consumers returning to your store (Figure 14-8). Even if the horticultural enterprise is purely interiorscaping with little to no walk-in clientele, for the few clients who do visit, the public areas of the company should be dressed for success. New varieties of plants should be present, providing accent to the interior. Show clients that the company practices what it purports to do; live plants in attractive containers improve interior surroundings and create an air of professionalism, style, and design.

Interiorscapers, from the technician to the company owner, should visit interiorscaped sites of all sizes. See the installation the way the client views it. Does it measure up to the highest standards or are there embarrassing features that need immediate improvement? It is important to visit sites maintained by competitors. If they do a good job of maintaining a site, it can be a useful learning tool in improving your company's services. If it is substandard, a professional, friendly conversation or letter to the decision makers could result in a new maintenance contract.

Sales Promotions

When starting a new business, you will search for business if you are hungry for income. Once your interiorscape firm has a foothold with an account, you can capitalize on selling additional products and services unlike any other provider.

Suggest to the account's decision makers that they give a floral or plant gift to each of their employees on their birthdays, provided by your company. The employer would need to provide the interiorscape firm with a list of employees and their birthdays. If the technician visits the site once a week on Thursdays and a birthday occurs earlier in the week, the technician could bring the birthday plant the week before. A 6-inch potted plant, of similar variety and size, could be delivered with a designated message. Care should be given to appropriately dress the plant in a decorative pot or waterproof basket.

Be sure to note that the plant is the employee's property to take home and enjoy. It may not be the best case for the plant to stay on site for a long period of time because it could revert to being cared for by the interiorscape firm. The cost of the plant could be added to the monthly bill.

Because so many interiorscapers provide Christmas and holiday decorations, having an established account provides the firm with a terrific inroad to suggest specific decorations (Figure 14-9). The interiorscaper would be responsible for installation, maintenance, and removal of décor. Early orders, where the client places an order in the month of August for delivery in December, can result in a percentage discount. Another take is to organize the client to place orders before major floral markets. For

FIGURE 14-9: Garland festoons and Christmas trees in traditional colors lend holiday spirit to this shopping mall.

example, the America's Mart Atlanta holds major buying shows in January and July every year. It would be smart for both the interiorscaper and client to think about holiday décor right after the holiday season, when the concepts are fresh in their minds. A walk-through of spaces needing design could be accomplished with thought to discarding worn decorations. With an order and, better yet, a cash deposit in hand, the interiorscaper could source and buy decorations that would be specific to the client's needs. This, coupled with a percentage discount due to an early order, creates a higher level of customer satisfaction.

It is always a good idea to provide incentives to drive sales, especially when an early commitment from the client saves you time and money. Interiorscapers should strongly consider setting and following a policy of providing discounts on full payment of annual maintenance contracts. The highest price could be placed on month-to-month contracts, with perhaps a small discount for 6-month contracts and a larger discount for 12- to 24-month contracts. If the plant-care provider is giving excellent service with quality products, it makes sense for the client to sign on for long-term care at a cost savings.

Parts of an Interiorscape Bid Submission

When supplying an interiorscape bid, several categories of information can be a part of the submission. It is always a good idea to offer a business card and contact information including cellular number and email and web site addresses. Provide the client with your company's mission statement and a brief write up of the purpose of your business. The submission should underscore the benefits of indoor plants, how they improve interior design aesthetics, soften architectural lines, and add a distinctively natural touch. Remind the potential client that good-looking plants aid in thematic development and reinforce corporate image. Plants lower stress and enhance productivity in the workplace.

Provide the client with images of spaces to be designed. If they are empty, show them empty. If already interiorscaped, show with old installation. Using freehand illustration or computer software images, contrast the old looks with new designs. Tell the client about the types of plants to be used with emphasis on design elements rather than on specific varieties. It is good to discuss various genera, but remember that availability may necessitate substitutions.

The core of the bid submission is the segment where costs are listed, ultimately arriving at a bottom-line fee. Installations are based upon the retail cost of materials, labor, and supplies. Maintenance fees are a separate issue because this service is billed to the client at regularly scheduled dates such as once a month. Maintenance fees are developed from estimating the amount of time it takes to water, dust, and otherwise finesse each plant in the account per maintenance visit. A 20-foot-tall tree would take longer to care for than a 10-inch-wide hanging basket. The total amount of time to service the account per visit should be compared to the cost charged for the horticulture technician's labor. The service fee should also take into account the amount of time and expense it will take for the technician to travel to and from the account.

As part of plant-care maintenance, the client must be informed that plants will occasionally require replacement. Some accounts will use plants that are more durable while other accounts will have a mixture of seasonal plants and dark green, long-lasting plants. With each service period, the client pays a small amount as an insurance to guarantee replacement of plants that have lost aesthetic appeal.

At this point, state payment terms in the contract so that the client fully understands when payment is due. Provide an address for remittance. It is recommended to charge the client interest on late payments and collection terms if you must sue for payment. Finally, payment terms are presented, and a tentative installation date is agreed upon. This portfolio may be presented as a hard copy, saved to a jump drive, or both.

Contracts

Agreements between entities such as a client and an interiorscaper deserve special attention, much more than just 5 to 10 minutes of conversation, some affirmative head-shaking, and a handshake. What good does that do? Neither party could possibly remember everything they spoke about let alone the unspoken assumptions made (Figure 14-10).

FIGURE 14-10: Contracts provide important understanding and reference in business transactions.

Contracts provide all parties with rules of conduct and are an important way of helping people to understand what will and will not occur in specific instances. There are many generic contracts available to small business owners today, but many of them are not specific to the needs of an interiorscape company. If a thorough contract is not used, costly mistakes could land a small interiorscape company in court and out of business. This is one of the many reasons why it is so important to gain experience in horticulture before starting a company. Practice and experience enables entrepreneurs to avoid disastrous pitfalls.

An important component of the contract packet is the list of plants included in care. Plants not recorded on the list are not a part of the contract and should not be serviced. It is important to provide a list of normal maintenance procedures in a contract. These activities include, but are not limited to, watering, dusting, removal of trash/debris, and trimming necrotic leaves. Since normal maintenance requires supplies such as watering cans or tanks, cleaning supplies, soap misters, and others, some accounts may require secured storage space be provided by the client. The client may also be asked to supply a safe ladder, which is consistently stored in the same place and is of appropriate height for servicing plants.

No plants are to be introduced into the interiorscape by any party other than the interiorscaper due to potentially introducing pests or diseases. Decorative containers must also be supplied by the interiorscaper based upon specifications from architects, interior designers, and the client. Containers are limited to the manufacturer's warrantee and no other warrantee can be made.

Any major actions not stipulated in the contract are not a part of the normal maintenance. For instance, your company will not be responsible for the cleaning or removal and replacement of very large specimens already in place. Perhaps your interiorscape firm is capable of cleaning the large palms just mentioned because you know you can hire short-term help. You may provide a statement that such services may be offered by your company, but are not a part of normal maintenance procedures. The interiorscaper should fully explain and define guaranteed replacement. When will the plants be replaced? What about the replacements? Replacements should be of similar size, look, and value.

It is a good idea to explain color rotations in a contract. Let the client know that color rotations involve special plants that are in bloom, have colorful, unique foliage, or both. These plants do not have as long of a display duration as foliage plants and must be replaced more frequently. Some color plants like bromeliads have bloom spikes that last for months, while more delicate plants like azaleas may last for only two weeks. The interiorscaper could categorize color rotations based on display duration, seasonality, or theme. For instance, a law firm might appreciate the sleek look of orchids or bromeliads rather than the high tea of *Hydrangea* or African violets.

Once the interiorscaper's services have been outlined, it is important to list the responsibilities of the client. One may feel that all the client needs to do is pay the bill and enjoy the plants, but the client should also be prepared to provide *guardianship* for them because plants are living organisms. Even though an employee of the client or another well-meaning person may feel compelled, no one but an agent of the interiorscape company should intervene and "care" for the plants. This may result in overwatered plants that become weak and susceptible to disease. Further, interiorscape technicians must always be provided with access to the plants within business hours. This may necessitate a security badge issued to the tech. Maintained plants should never be moved. Plants that have been moved because they are deemed to be in the way, even temporarily, may suffer due to even slight changes in light intensity. It may take months for them to acclimate, if they adapt at all.

All plants and containers should be protected from damage caused by carts or machinery. For instance, a forklift operated by a third party installing special banners in a mall accidentally toppled a planter, breaking *Dracaena* canes and poking holes in the decorative planter. The planter and plant can be replaced in a reasonable amount of time, but it is not the responsibility of the interiorscape company to cover replacement costs. Those costs should be funded by the client, who hopefully has been reimbursed by the third party who caused the accident.

The client should also be responsible for vandalism and theft. Sometimes, guests at hospitality sites such as resorts or luxury hotels feel entitled to objects that are not a part of the "package" and may steal pillows, robes, and bromeliads. Unsupervised children can provide their own brand of vandalism. From time to time, unusual things can happen. Thankfully, these occasions are rare.

What would happen if water were shut off in the building? This could be the case if a major renovation was to occur. Plants will still need water during the interim period and the client should be responsible for its cost. Temperature, too, is important. The interior environment must not fall below 55°F. Plants should not be exposed to fumes such as exhaust, cleaning chemicals, or other compounds that could cause harm. They should not be used as drains for the disposal of coffee, liquor, or other liquids.

Contract duration should also be listed, whether monthly, yearly, or biennially. Payment terms should also be given. It is also a good idea to charge additional fees for late payments. What would happen if a client did not pay after an extended period of time? It does happen and may provide a reason for termination of services. The interiorscaper should also leave some room for price changes in plant care at the end of the contract period, sometimes upward when costs rise or sometimes downward in the case of added competition.

All parties must agree to all terms of the contract, and this should be stated, followed by signature and date lines. A contract draft should be reviewed by a lawyer, who would add additional, valuable information. It is advisable to speak with a representative of a law firm to get an estimate on the cost of such services. It may seem overwhelming, even considering you have listed much of the content, but keep in mind that an investment in a contract model will remain durable and usable for many years.

Honesty and clarity between the interiorscape company and its clients will consistently pave the way for mutually rewarding relationships. The interiorscaper must never promise something he or she cannot deliver. After the sales and installation are complete, it is important to keep in contact with clients. Technicians are in and out for regular maintenance, but this is not the entirety of plant care. Every few months or so, an interiorscape management staff member should visit client sites and decision makers to determine whether they are satisfied. As horticultural entrepreneurs, we must ask ourselves "What does it take for our clients to be delighted about our products and services?"

SUMMARY

When studying interior plants, savor the information you learn. Even the smallest bits of data you remember may very well be used in future interactions with clients and staff. Knowledge is the foundation of building a great interiorscape business. Clients value working with a plant expert rather than with someone with a lower level of understanding. Successful interiorscapers gain experience through entry-level jobs in horticulture before they start their own commercial enterprises. Interiorscape maintenance companies do not require much initial capital, but they do require product sourcing experience, plant care knowledge, and the ability to communicate and share the importance of plants in the interior with potential and existing clientele.

CHAPTER 15

Techniques

INTRODUCTION

People love horticulture as a pastime and as a career because it is engaging. It not only involves theory, seeking to answer the questions of how something works, but it also involves activity. Once your hands and head are working with plants, the rest of the world seems to fade away. Any worries or concerns are gone, at least temporarily, thus making horticultural work great for calming the nerves and helping someone to gain perspective.

Being able to learn with your hands and keeping active is fun. Many students remark they like learning by doing, working with their hands. The best horticulturists with the most successful gardens, inside or out, are those with many hands-on experiences over many years.

To be a good horticulturist, you must practice horticulture. A plantscape practitioner practices the techniques of indoor plant care, the way to carry out particular tasks, with numerous plant materials in a multitude of settings. This text helps people to understand the underpinnings of why we do what we do.

In order to get started, several tasks associated with plant care are outlined in this chapter. To many people, this chapter may be the most important because it explains how to accomplish many of the duties associated with the care and handling of interior plants.

GREEN TIP

Talk about plants as often as possible. Use the topic to initiate a conversation and to educate people about plants.

Specifying a Job

At one point or another, it will be necessary to meet with corporate decision makers to discuss their interiorscape needs. It is important to remember that during this process, the interiorscaper should listen and observe much more than speak. This is the point in time in which we gather information about the spoken and unspoken needs of the company for an interiorscape (Figure 15-1).

Remember that corporate clients are asking you to create a beautiful interiorscape for them based upon available budgets. The clients may be very savvy and know a lot about what they do, for instance, consulting, law, or medicine, but they have a need and desire for what plants can do.

Private clients may know a great deal about interior plants, their names, care, and stylish display, but they are asking you to take the helm. Do not get intimidated by displays of fabulous interior design or depth of knowledge by your potential client.

It is easy to fall into the trap of saying too much. In an effort to communicate to the client that they know what they are doing, some "green" interiorscapers, in this sense meaning "new," say too much, painting a colorful picture of what they can do with massive, impressive plants and over-the-top displays. This is fine if the final installation meets expectations, but it is far better to exceed client expectations than to fall short.

Before a visit with potential clients, it is a good idea to tour the site. Get a feeling of what would be best for the site in terms of proportions, plants, and possible limitations when you do not have the distraction of interacting with other people. If the site is an office building, you might ask for a pre-consultation tour from the client or a

FIGURE 15-1: Part of being an effective business person is being a good listener.

representative in order to gain security clearance. You might make a bad impression if you take pictures or make sketches of a space without the consent to do so. After following all the proper procedures to gain access, an initial site visit will offer a valuable sense of space and show the potential client you care about the job.

Interiorscape sales and design staff should create a checklist that aids in the initial specification process. If clients are present during an initial walk-through of a space to be interiorscaped, distractions can cause staff to forget to check on the most basic things.

Interiorscape staff should ask for permission to take digital images of the space. Pictures can easily be taken on mobile phones and uploaded to digital files. As the bid submission progresses, images may prove invaluable to answer specifications.

What is the available light? If it is natural, it is the best for live plants. Artificial/supplemental light may help in sustaining plant display life for longer periods. Decorative lighting does not help much due to limited spectrum and intensity. Take care not to block or place plants too close to these sources. The effects of sunlight and temperatures change during the seasons. What are the anticipated seasonal differences based upon window and door placements, window shading, and coverings?

Where are the plumbing sources in proximity to plants? If watering, cleaning, and other equipment and supplies must be kept on site, where is the storage space? Is its size adequate and is it secured?

Are there sources of air movement or air currents that cause drafts detrimental to plant displays and plant health? Temperatures can fluctuate in microclimates near exits, at night, or on weekends when systems automatically lower or raise thermostats to save energy.

FIGURE 15-2: Focal areas can be emphasized and created. In this room, the fireplace is highlighted, but additional plant placements could be made throughout the room to provide dramatic accents of greenery.

Placement Possibilities

It is a good idea to highlight focal areas of sites to be interiorscaped with plants. While you are at the site, note the places where your eyes are drawn because these are the focal areas of the space. Remember, too, that focal areas can be created with the use of indoor plants. Find new clients and create win-win situations, making their sites appear distinguished by your excellent plant placement and maintenance services (Figure 15-2).

If the building has security services, you will need to know what the requirements will be in order to clear security for regular maintenance. Do they require a photo badge for every person from your company who will provide horticultural services? Technicians may only have to report to a security officer upon arrival at the site. What are the expectations of building security staff at the time of installation?

Take some time to observe the foot traffic patterns in lobbies and busy areas of sites to be interiorscaped. Seeing how a space is used will provide valuable information about where plants should and should not go. Do traffic patterns change during the day or night? Do the types of people traveling through the space change? For example, a hotel lobby may have increased traffic in the late morning and early afternoon. At night, the hotel's center space, which may seem open and airy in the day, could be full of convention revelers patronizing a lobby bar.

Bear in mind where plants and supplies will be delivered. A loading dock for plant installation and major servicing is necessary to bring in large-scale plants and plants in quantity. Your delivery vehicle must be able to access that area. What time will the installation occur? Depending on the time of installation, the loading dock may be highly congested or perhaps closed to deliveries during employee breaks.

Companies hiring interiorscapers depend on their clients for revenue generation. Your customers want to treat their customers with great care and respect, so all aspects of interior plant services must follow these philosophies. In order to take care of the plants yet stay out of the way, find out the best times of the day and days of the week for servicing indoor plants. There is an art to making work appear effortless.

When technicians arrive for plant services, where should they park? This may be obvious with some sites, but may be a sticky subject with busy, downtown sites or government property. There is great competition for service parking in congested, urban areas. Interiorscape techs deserve reasonable access to parking spaces in order to tend to perishable interior plants.

Decor

When viewing the space to be interiorscaped, note the height of the room. There is great potential to create "wow" reactions to the placement of interior plants when larger materials are used. Recall the use of the rule of thirds with interior plant placement. If you want a plant to measure 2/3 the height of the room, what is the approximate height of the space? This refers to the design principle of scale. It is a reason why it is important for interiorscape sales staff to have a working knowledge and vocabulary in design principles. It is indeed what they are selling.

Another design principle—proportion—asks the horticultural designer to consider the amount of foliage plants a space can accommodate. What will dominate in the room, negative space or plants? This is often dictated by budget, but designers should keep in mind the theme and use of the space. It is a good technique to provide clients with more than what you think they will want, and then adjust the proposal downward to meet budgetary and space needs. This practice is often used as a sales technique, and clients will communicate budget restraints. It is nearly impossible to work in the opposite direction, however. If a plantscape salesperson is timid in terms of plant placement, suggesting just a touch of greenery, clients will either accept the contract with a miniscule number of plants or move on to a plantscaping supplier they perceive as having better ideas and products.

Note the color palette of the interior design. Interiorscapers have the ability to suggest, sell, and install color rotations highlighting important areas of an interior site. With the use of color-combining theory, it would be spectacular to make use of color-opposites to help an interior come to life. Even with installations that are low-key, it is possible to use plants with varying tints and shades of green along with variegated patterns to please a client.

Note artwork, fabrics, ambient music, and the many design elements that come together to provide a theme and feeling of a space (Figure 15-3). The design elements can be challenging to put into words, but the interiorscaper can effectively support and add to the feeling of a space with well-chosen plants. Experience is the best teacher when it comes to building upon an established theme. Interior designers can offer specifications, but the plantscaper knows the horticultural requirement for plants indoors. These two professionals working together can create a setting where the outcome is greater than the sum of the parts.

FIGURE 15-3: This hallway in front of a meeting space suggests color combinations and design possibilities for interior plants and their furnishings.

General Light Guidelines

There can be a wide variety of light intensities indoors. Table 15-1 provides a bit of insight into what the light might be like in a few different situations.

When an abundance of light is available indoors, it is possible to display and grow numerous interior plants. Usually, this is not the case. Nevertheless, several types of plants withstand very low light levels and keep their good looks for months, even years. Some examples are listed in Table 15-2.

Certain characteristics of plants provide helpful clues for their suitability to given light levels and care. Table 15-3 provides generalized data helpful to the interior plantscape technician. It is possible to memorize the names of such plants, and certainly, after working in the industry for a few years, you will have a working knowledge of what the site requires and what you can source.

TABLE 15-1: Typical Light Measurements

A dark corner	10–30 footcandles
Office	50–300 footcandles
South-facing window	2000+ footcandles

TABLE 15-2: Minimum Tolerance Light Levels

Ficus binnendijkii 'Amstel King'	150 footcandles
Chamaedorea	50 footcandles
Aspidistra	50 footcandles

TABLE 15-3: Characteristics of Plants for Low Light and High Light Areas

Plants That Do Well in Low Light	Plants That Do Well in Brighter Light
Thick and leathery leaves	Thin leaves
Dark green leaves	Light green leaves
Slow growers	Faster growers
Less water	A bit more water
Slow to wilt	Quicker to wilt
Moisture stress at petiole	Moisture stress in color
Larger but fewer roots	Many fine roots
Less acclimation shedding	More acclimation shedding
Lose older leaves first	Lose interior-positioned leaves first
Acclimation easier	Acclimation more dramatic
Adapt to coolness better	Not as adaptable to cool temperatures

GREEN TIP

Indoor plants aid in reducing volatile organic compounds. Because they improve air quality, rethink the number and ways more plants can be introduced into an interior design. New container designs allow plants to hang from ceilings or walls, which will free floor space in a home or office.

When a plant stops producing new shoots, intermodal spacing becomes elongated, and stems become thinner, it is surviving at its minimum light tolerance level.

For a detailed listing of indoor plant materials, including appearance, pot sizes, and light requirements, see the newest edition of *The Guide to Interior Landscape Specifications* by ALCA, the Associated Landscape Contractors of America.

Plant Presentation Techniques

Some of the ways interior plants are "planted" in displays may seem strange to those used to exterior landscapes and gardening. The two spheres share similarities, yet there are many differences, too. Interior plants must be displayed in ways that support plant life and work with interior design.

Plants Displayed in Cache Pot

A live plant can be displayed where its "grow pot," its original pot from production, is placed directly into an ornamental container, referred to as a jardinière or cache pot (sounds like "cash poe"). Keeping the grow pot intact means that the entire plant is easily removed when necessary for replacement. This practice also ensures drainage from the plant, but may also create a reservoir of the leachate in which the grow pot will sit.

Prior to adding the grow pot to the jardinière, a strip of foam rubber can be glued to the jardinière's inner side for stability, or it can be fit inside without gluing if the fit is snug. This protective strip forms a barrier between the grow pot and the decorative pot so that mulch will not fall down in the space between the two pots.

Many interiorscapers use a hard, plastic liner between the grow pot and jardinière to catch leachate. With porous containers like unglazed pottery, a liner layer will protect the decorative pot from collected water which would cause unsightly water and soluble salt stains. As always, techs should monitor plant saucers and remove them from standing water. It is harmful to roots and may aid in causing odors and conditions that support shore flies and fungus gnats.

Direct Planting in Beds and Large Planters

Some plants may be directly planted into larger beds and planters. When this is done, care should be taken to gently loosen hardpan on the surface of the plant and to gently tease roots openly growing on the outer sides of the root ball. Insert the plant into the new media at the same height it was previously growing. Follow this with a thorough watering that soaks the root ball and surrounding soil well. Finish with a layer of mulch.

Grow Pots Sunken into Beds and Large Planters

This method uses techniques borrowed from the previous two methods (Figure 15-4). Large planters and planter beds are pre-filled with media. Selected media should be light and kept unwatered. An installation crew makes holes in the media where plants will be placed. The plants chosen for display are kept in their grow pots and sunken into the larger planter media while some of the media is swept upon the grow pot rim to aid in concealing it. A layer of mulch covers any remaining portions of the pot. Base planting, which can also be installed in the same fashion, helps in concealing pots, especially vining-type plants like *Epipremnum aureum*, Pothos, and *Philodenron*.

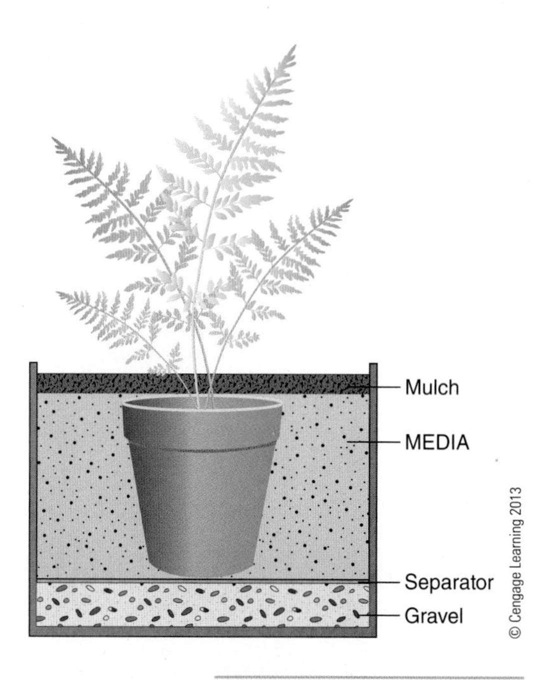

FIGURE 15-4: Plants are kept within their grow pots and sunken into lightweight media filling larger planters or in-floor beds.

Keeping plants in their grow pots helps in quick removal of individual plants if necessary. It also cuts down on the amount of water necessary to sustain plants and less water means less weight of the overall planter. This could be helpful when trying to turn the planter to alleviate the effects of plants growing toward light sources.

Subirrigation Systems

Many interiorscapers enjoy the success of keeping indoor plants within subirrigation systems. Subirrigation refers to the dispersion of water under the soil line rather than top-watering plants. The use of subirrigation to keep plants watered is ancient. Early cultures utilized unglazed earthenware jugs buried into the ground with only the lip or a spout exposed (Figure 15-5). Plants were planted next to the jug and their roots flourished, receiving water through **capillary action**.

In the world of interior plantscaping, many materials and methods have been employed to facilitate movement of water from a reservoir to roots. The processes are different from those used to water exterior turf and gardens; indeed, they are much simpler. Obviously, soil cannot remain saturated. Plant roots need air-filled as well as water-filled pores surrounding them. Constant water saturation eliminates oxygen in the root environment and can ultimately kill the entire plant. Good subirrigation systems must allow for appropriate balance of water and air in the root environment.

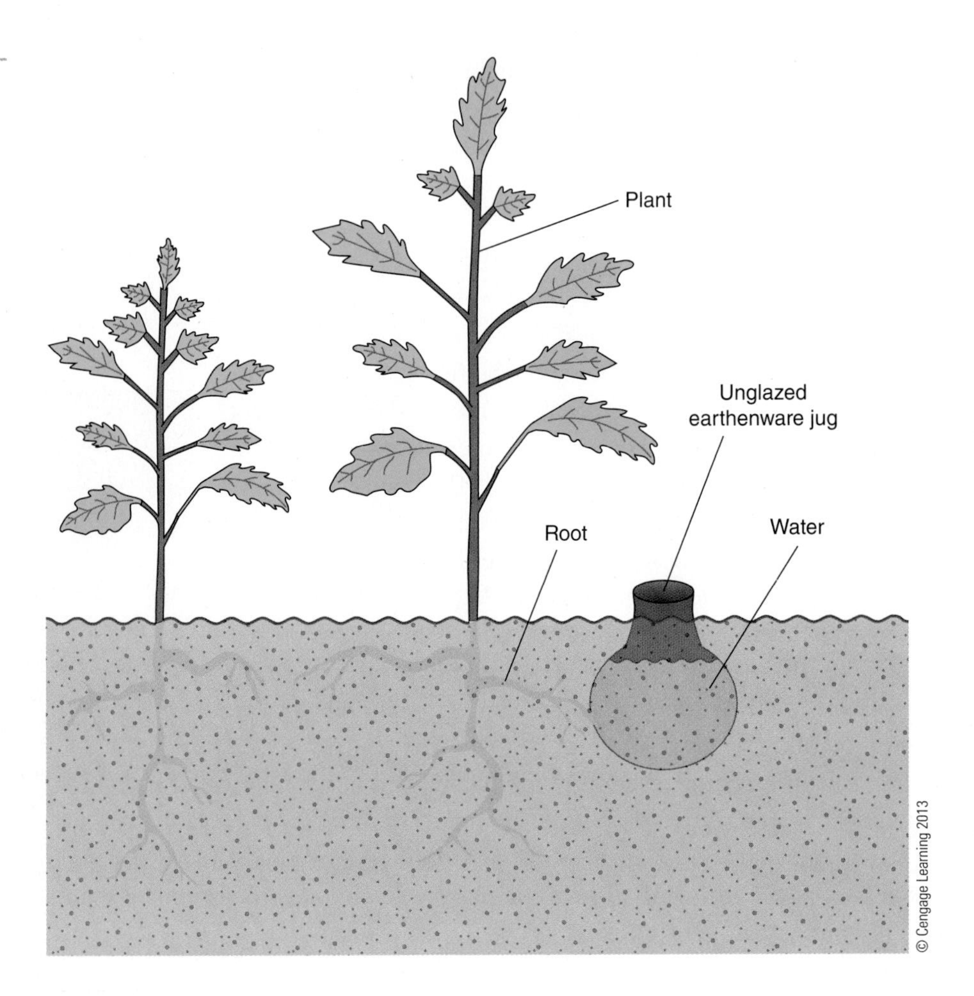

FIGURE 15-5: Unglazed earthenware jugs slowly release water in soil.

A variety of subirrigation systems exist in today's marketplace, many of which are directed toward the interiorscape trade. Some systems use a series of hefty, hard plastic casings, which act as the underground reservoir. At different points within this tube-like system, there are screen-covered, plug-like indentations. At installation, the casing set is assembled to follow the contour of the planter, whether round or rectilinear, and positioned below the root zone of plants. Then, a vertical fill-tube is attached to the horizontal casing, which allows water to be added to the reservoir. Next, soil is added, with some worked into the system's indentations. Soil follows the contours of the plugs and is in contact with water. More soil is added, and then plants are directly installed. Capillary action from these underground reservoirs allows for slow uptake of moisture through the root environment.

Another type of system uses a vacuum effect to keep plants watered over weeks or months. Instead of a series of connected pipe reservoirs,

at the upper end of the column. This is due to the collection of more water-filled pore spaces at the bottom of the wall.

Once the wall is in place and irrigation is operational, the next step is plant installation. Pots are removed from the plants and media is rinsed off the roots. Using a utility knife, cuts are made into the foam or fabric layer and the plants are then directly planted into the substrate. Plants selected for the interior green wall must be appropriate for light levels present. It should be noted that light intensities could be different at the top and bottom ends of the wall due to skylights, artificial lighting, and shadows.

Because this type of green wall is more permanent, thought must be given to the patterns made by plants and how plants may grow and change the pattern, for better or for worse. Large green walls use corporate logos and take careful planning in terms of scale and proportion relating to the crispness of the logo, its color, and contrast. It must be perfect because it reflects a trademarked corporate logo, hence the company's image.

FIGURE 15-8: An established type of green wall.

Simple patterns such as crescent-shaped swaths of one type of plant on a background of less-innocuous plants may provide design inspiration. Another approach may be more like a crazy quilt, mixing plants about the wall with no particular pattern, only regarding the light requirements of selected plants. In time, as plants become accustomed to their new environment, leaves will turn toward the light and stems will spread outward creating a more natural look. Plants will begin to adapt to their new surroundings (Figure 15-8).

In the planning of green walls, available light must be regarded carefully. If the space offers low light, then the plants will grow very slowly. The wall will not fill-in completely with greenery in a short amount of time. It may perhaps take years. This could be a problem if the substrate is not very decorative, such as using a gray felt layer. On the other hand, if the substrate is more natural or otherwise attractive, greater amounts of negative space could help plants maintain their own personality rather than being "just one of the crowd." This underscores the importance of knowing the expectations of the client.

Just about any type of plant can be cultured this way. In order to make decisions about what to use, remember the basics. All a plant needs is water, nutrients, and light. Consider plant growth habits in developing the design on paper or as a model first.

Green walls are focal points. If this green focal point bears unusual, artistic, well-designed patterns or corporate logos, it will be seen constantly and scrutinized by decision makers. It must be in top shape at all times with no dead plants, foliage, or unsightly problems. Many interiorscapers who specialize in green walls state that imperfections are more visible and thus require immediate attention. This should be factored in when accepting a maintenance contract.

Imagine a restaurant with a wall of herb plants. From a thematic standpoint, this could generate marketing buzz for the restaurant. Imagine the interest of customers who would experience the sight and fragrance of culinary herbs and the conversations it would start, inside the restaurant and long after the dining experience, as part of word of mouth advertising. Most herbs are light-loving plants and would not tolerate the low light levels of indoor restaurants very long. They would need to be maintained and replaced more frequently, increasing monthly maintenance charges for plant care and alternates. Though it may cost a bit more to use herbs rather than low-light, tropical plants, do not underestimate interest in the new and novel as well as the pocketbook of your potential clients.

Interior horticulture crosses over into many other specialties. Tackling a major project such as a three-story green wall requires the expertise and skill of several designers and trades people. You do not have to be an irrigation specialist in order to design, sell, and maintain specialized installations. You must possess the skills to source and maintain good business relationships with other professionals.

As a complement to vertical plantings, interiorscapers can take a refreshed view of hanging plants as a form of plant presentation. They can be single-specimens or mixed genera using upright, shrubby, and cascading-type growth patterns. Hardware for creating a hanging planter can be imaginative and quite decorative. Interiorscapers should not feel limited to using only commercially manufactured hanging plant containers, but they are cost-efficient because the product is finished and ready for use.

It is always possible for a horticulturist to design a display mechanism and then have it fabricated by persons who specialize in the medium. For example, a sheet metal worker could fabricate a steel hanging planter of approximately 4 feet in diameter with the appearance of a wave. Frequently, all these artisans need are a drawing, specific measurements and, of course, payment. The finished planter will result in a one-of-a kind, site-specific piece of horticultural art. Who better than a horticulturist to design an object like this, fully intended for the culture and display of living plants? This is the spirit behind the green wall theory. At some point, someone asked how he or she might possibly use indoor plants in a unique and different way.

Maintenance Procedures

Once an installation is completed, it is handed off to interiorscape maintenance staff. Most companies assign one technician per client site. In this way, the technician prides herself on the appearance of the plants. Clients recognize their technician and, in larger office buildings, it is faster for them to clear security. The technician knows the ins and outs of the site, quickly locating parking, water sources, and trash disposal. She will know the appropriate time to visit the site and will stay on top of problems and solutions. Even with these benefits, it is valuable if supervisors visit client sites from time to time, unannounced. They can evaluate the quality of the plants and the technician's work, and they help to provide a second pair of eyes to identify problems, both entrenched and potential.

Watering Techniques

Adding water to the soil of a potted plant not only benefits the plant, it benefits people, too. Moisture from soil evaporates, gently humidifying the air and thus improving indoor air quality. This is especially helpful during the dry, winter months.

Watering plants takes time and should not be rushed. Leach watering is the most popular form of watering interiorscapes, but it can be messy and time-consuming. Remove leached water from saucers and jardinières when possible. Within the water and air relationship, the downward movement of water in the soil column helps to draw air downward. Thus, water-filled pore spaces alternate to become air-filled pore spaces, necessary for root health.

As plants acclimate to lower light intensities, they will gradually need less and less water. This is a positive sign of plant survival. If a space is cool, plants will need less watering. A temporary reduction in indoor temperatures, such as during a long holiday, may result in plants not needing as much water after workers return and temperatures are raised. Plants in reception areas and vestibules that are more constantly cool all the time do not necessitate frequent watering. When placing plants near a window, provide space between leaves and glass so that they do not come in contact with each other. High light intensities may "burn" the leaves.

One of the most helpful tips to solve the dilemma of knowing when to water is called the "finger test." For plants grown in pots from 6 to 22 inches in diameter, simply sink your index finger into the pot for a depth of about 1 inch. Upon removal, if any soil particles cling to your finger, the media has enough moisture content to sustain the plant for another day or perhaps more. If very few to no particles cling to your finger, give the plant a thorough watering so that some water flows from the holes of the grow pot.

Water the soil surface evenly if the plant is being hand-irrigated from the top of the plant. Add water so that the soil ball is thoroughly saturated and that a small amount of water percolates through the drainage holes at the bottom of the pot. This is not necessarily tidy work, though. Some plant caretakers will add small amounts of water to a pot to avoid collection of excess water in the plant's saucer, but the better way is to add water until excess leachate runs from the drainage holes at the base of the pot. Once the water slows or stops, empty the saucer of excess leachate.

The problem associated with too small of a quantity of water is that roots will live only when they are nourished with water, followed by air. Consistently dry areas within the soil will not provide an environment for roots to flourish. As the foliage portion of the plant develops, the plant could grow to be top-heavy and lose physical stability.

The symptoms of overwatering are quite similar to those of underwatering because the problem is in the root zone. Recall that too much water lessens air space, reducing respiration and ultimately resulting in root death. Too little water keeps delicate root hairs in a dry environment resulting in their death.

After a period of time on the job, plant techs gain a sixth sense with their plants. Good techs can sense when plants will need water before they even see them. They are able to build this ability, more than likely experiential rather than the paranormal, into their maintenance schedules. A plant may register as moist and it may feel and appear moist to the touch, but this must be weighed against the period of time until the next maintenance visit. Will the plant show signs of its stress, detracting from a healthy look? It may be better to water the plant this week, providing a little less water than usual, and skip watering it the following week to help keep it on a regular schedule. Try to learn each plant's wilting point—the way the leaves appear, the overall plant's appearance, about how long it takes to get to that point, and what a moisture meter measures at that point.

Many techs who are new on the job or working in interiorscapes that are new to them have found success with moisture meters. The method of using them is rather simple. Insert the probe tip toward the middle of the root ball rather than inserting it only on the sides of the pot between the pot's wall and the root ball. This will give a more accurate reading of the root zones having more active water uptake. It may seem that the probe stick of the meter could damage roots, but unless driven into main taproot, moisture probes usually do not hit roots but go around them. Probe "sticks" in the soil actually help the root environment because they help aerate the soil.

They are inexpensive tools and do not last forever. Meters may need to be replaced after 6 months of professional use. They are somewhat personally calibrated, meaning you learn the plant's needs over time and adjust to what the meter reads. Most plants need to dry somewhat between watering while others could become too stressed or even die from this type of treatment. Moisture meters are great for new technicians and help them to get to know plants and planters in individual accounts. In time, as you learn more about plants, you will not need a moisture meter.

Moisture meters work with most soil types except lava rock media, which is often used with plants grown in Hawaii such as *Howea* and *Anthurium*. When a probe is inserted into ground lava rock, it always registers as "dry" even though the media was saturated.

Repotting

Interior plants usually do not need to be repotted frequently. Since light levels are low, interior plants do not develop new leaves quickly; therefore, they are not developing roots to support the plant's needs for water and nutrients. A tropical plant kept indoors can stay in the same pot for years.

An indoor plant can benefit from having its soil refreshed after a few years of display. Remove the plant from its grow pot. Discard any loose medium that is not held in place by the plant's roots. Very lightly tease away soil clinging to the root ball and discard it. Wash and dry the pot, then add a few inches of fresh media to the bottom of the pot so that the root ball is staged at its previous height. Add fresh media to the sides of the root ball, gently tamping it down, then a bit on top. The plant's roots will respond to this soil refreshment with new root growth, replacing roots that have naturally died over time. The added benefit of this repotting is seen in the replacement of overspent soil with nutrient-laden soil along with a greater amount of pore spaces for water percolation. If a plant is in a growth-type situation, repeat the same steps as above, but use a pot slightly larger in diameter than the previous. It is best to move up just a few inches at a time to keep the plant proportional to the container.

Cleaning Plants

Leaf cleaning removes fertilizer and pesticide residues accumulated in commercial production. Layers of residue often leave a whitish coating that should be removed. Technicians should be instructed to clean leaves, using a commercial leaf cleaner, about twice per year or more if needed. It is often quite striking to see just how much grime can accumulate on interior plant leaves. Outdoors, leaves are continually cleaned by wind and rain, but indoors, dust settles and can even adhere if cooking grease is in the air.

Some plants do not respond very well to shines and would stay healthy if they were used at weaker strengths. As always, it is best to follow manufacturer's recommendations and test one or two leaves on a plant and observe the results over a period of one or two weeks. In the mean time, a weak solution of liquid hand soap and water will remove grime and leave the natural shine characteristic of many tropical, thick-cuticle leaves. Some leaf cleaners leave a shiny, highly reflective appearance while others leave a softer shine. If anything, it is always good to start with naturally clean leaves where the natural shine of the leaf cuticle provides the glow.

When technicians apply commercial leaf cleaners, they should make applications to the top sides of leaves, not the undersides. Remember that plant stomates are mostly located on leaf undersides. Leaf shines may coat stomatal pores, thus not allowing them to fully exchange gasses between the interior leaf environment and the open air. This could ultimately cause leaf yellowing and death.

Numerous materials can be used to clean a plant. Soft sponges can be cut into smaller sizes with scissors and are able to clean small- and large-sized leaves. Soft fabric mitts are available from interiorscape suppliers and can be worn on both hands to gently draw off accumulated dust and grime. It is best to support the underside of a leaf with one hand while wiping away dust and dirt with another hand. Rinse sponges or cloths in soapy water solution. When hand-applying leaf cleaners, it is important to be very gentle, as if you are applying sunscreen on a baby.

It takes care and a lot of patience to properly clean a larger plant, especially those with many small leaves. For example, a 7-foot-tall *Ficus benjamina* will take a plant care technician about 40 to 60 minutes to fully clean all leaves. When cleaning a plant on site, always take care to avoid creating a hazard. People could fall and injure themselves. Avoid water mists getting on walls, floors, and furniture. If water mist or splash gets on a surface, wipe it dry with a clean cloth. Techs should always have access to clean cloths in order to dry surfaces quickly and efficiently.

Periodic dusting with a feather duster is also widely used to knock off very light, dry dust on leaf surfaces without much contact. This is not always the best way to clean because heavy dust is merely scattered to other interior surfaces.

Infestations and Infections

As our larger industry becomes more conscious of the long-term effects of pesticides through harm to other species and resistance, many interiorscapers are arriving at the same conclusion: Don't use them unless you absolutely must. Keeping control of insect pests and disease starts with buying indoor plants that are healthy and right for the space.

Indoor plant pests really do not bother people much except for species that fly about such as shore flies, fungus gnats, and whiteflies. Even so, these creatures are not biters and really do not care about mixing with humans. Piercing, sucking insects such as mealybugs and aphids secrete honeydew, which can make a floor or other surface sticky. Clients may think a light stickiness is caused by plant sap and easily dismiss it. The secondary problem of black, sooty mold may not easily be seen.

Scouting is part of the cleaning process as part of regular maintenance that the client hires us to perform. Most of the time, plants will not need lengthy maintenance visits, but when insects are present, time must be taken to manually remove them. As long as they are not too bothersome to our clients and their guests, and as long as they are not harming plants too much, we can learn to live with infestations. After all, insects live freely outdoors. The key is to keep populations in check. Use cotton-tipped swabs to reach between minute leaf axils. Place used swabs and other disposables in closed paper bags, then take them to a dumpster or vehicle for disposal. A good technician does not draw undue attention to an infestation. It is not necessarily understood by every guest and employee that insects have important roles in nature. They stereotype them as being signs of poor housekeeping, spoiling food, or decay.

It may take longer periods of time over months to lower the populations of larger infestations. You can remove them with strong water blasts appropriate to the size of the plant, from a plant mister or from the nozzle of a hose for larger trees. It may need to be accomplished out-of-doors.

Diseases on indoor plants are common when plants are stressed, usually due to overwatering. Knowing how and when to water a plant is the strongest line of defense against plant diseases. Specifying and installing the right plants for the provided light and temperature levels underscore that interior plantscaping is a site-specific decorative art.

Installation Procedures

The most precious commodity at installation is time, so it stands to reason that adequate planning will aid the process. Organize the plant order to arrive about one to two weeks prior to installation if possible. This will allow new plants to be quarantined before their introduction into existing interiorscapes. It will also give time for media to be thoroughly leached, removing excess fertilizer salts that probably are not needed in their new location. Take care to cut away necrotic (dead) tissue or yellow tissue. It should not be present on new plants, but it is good practice to check on this. Try not to cut away healthy, green tissue, as it will eventually brown. A leaf or two may be damaged or have died during shipment. This is not unusual and is easily remedied with a pair of scissors or clippers. This small task is much better accomplished at the interiorscape company office than at the installation site where people may misunderstand the activity.

Many interiorscapers remove older leaves on plants prior to installation. The older leaves are the first to yellow and die. In order to save a bit of maintenance down the road, but perhaps more importantly, to avoid customer complaints about "dying plants," this can be a good practice. Many clients do not like to see plants trimmed. They do not understand the importance of an overall healthy look and feel as though the technician is "hurting" the plant. Such preparatory maintenance is a very good idea. It is best to keep trimming to a minimum if being done in front of clients. It is better to clean plant leaves in front of people rather than anything that may appear to be "hurting" a plant, even though we know much better.

Hours before the delivery vehicle is loaded, it pays to re-wrap plants to avoid damage during loading, shipping, and unloading. If possible, save the plastic or paper sleeves in which plants were originally shipped and reuse them for their transport to the installation site. Take special care with new, fragile growth. It is far too easy to break terminal growth, which can ruin the aesthetics of the plant.

Load a van carefully with the full expectation that sudden stops or swerves will occur. When arriving on site, unload plants as soon as possible, preferably all at the same time and all at the same place. This will aid in gaining speed and keeping installers organized. It will help techs to monitor all the plants and not lose track of them prior to installation.

Devise a time schedule and map out delivery locations within the building for the installation. Do not rush, for indeed haste makes waste. You want to make sure that everyone who sees the installation observes professional staff members who know what they are doing. It is unpleasant to see a scattered installation, with a crew that does not seem to know what to do. It makes a client uncomfortable.

Know the delivery route, both from the origin to the installation site, and once indoors, the loading site (dock, doorway), initial holding site (which should be near the installation), freight elevators, and water sources (you should find these when inspecting the site to make a bid).

Take care when setting down a hanging basket. Vining material will be crushed or otherwise damaged when under the pot. Keep such plants sleeved or, if unwrapped, high on tables so that the vines are not stepped upon.

It is important for a sales staff member or crew leader to converse with people at the account as the installation commences. An interiorscape company should have a friendly brand, not abrupt and certainly not unfriendly. On regular maintenance visits, it is not wrong for a horticulture technician to sustain light conversation for a few minutes, especially with receptionists and office managers, at the end of the maintenance visit. These people are influential in a company's decision to hire you, or even to have indoor plants present.

Pruning

From time to time, it may be necessary to prune plant stems in order to improve and maintain the look of an interior tree. Plants may require pruning because they are in "growth" situations, the light intensities are bright enough, and light duration is long enough so that plants actually grow. They may, however be disproportionate to their display containers. There may be renegade branching where one or two stems stick out excessively from the rest of the foliage mass as in the case of *Ficus benjamina*, *F. maclellandii* 'Alii', or *Bucida* (Figure 15-9). When pruning larger limbs, always employ

FIGURE 15-9: It is sometimes necessary to prune large interior trees in growth-type lighting. Removal of some **codominant leaders** allows remaining leaders to establish as central leaders, improving plant aesthetics in the long term.

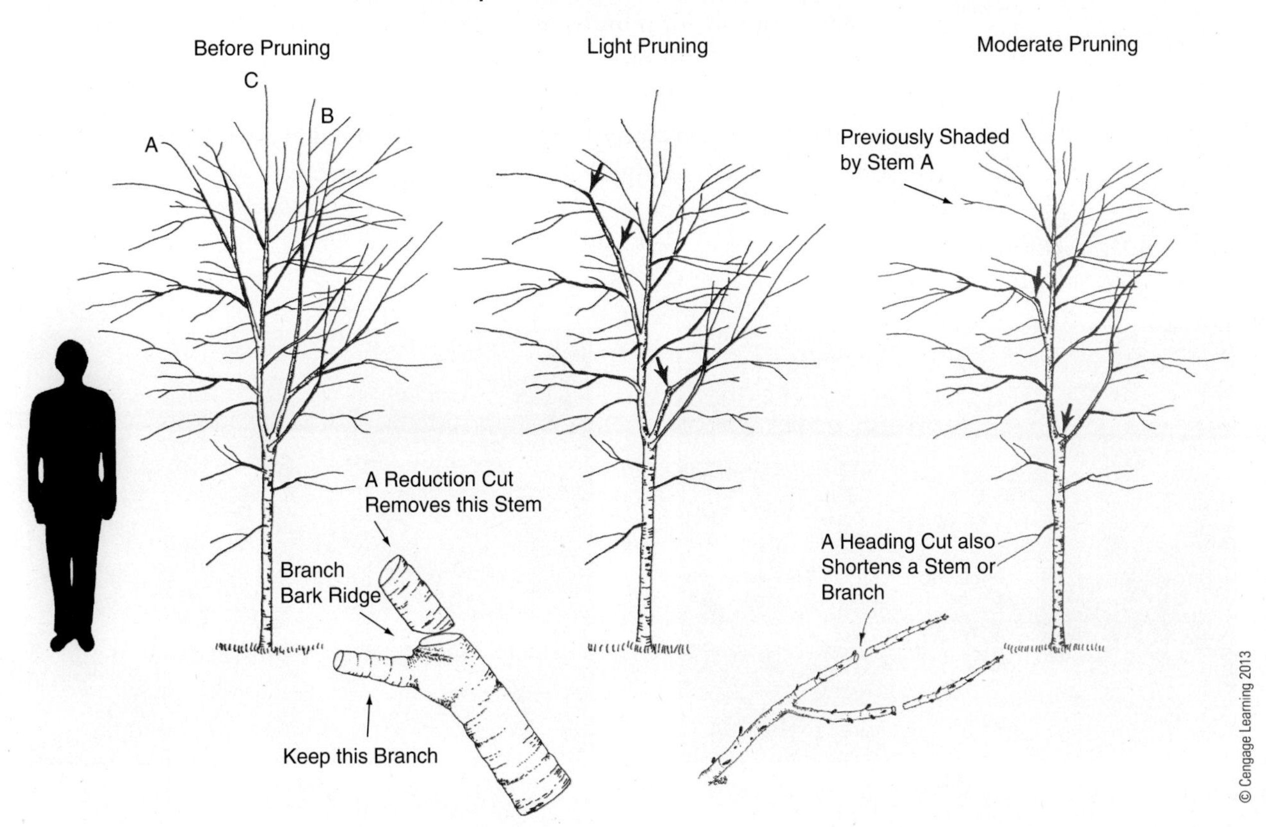

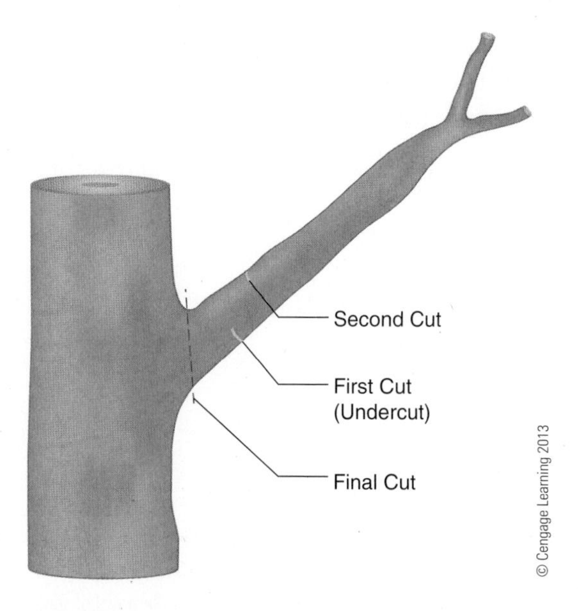

FIGURE 15-10: In the first cut of the three-cut method, saw the limb about one-third through, until the blade begins to bind. The second cut is further out, to remove the weighty branch. The third cut removes the stub.

the three-cut method (Figure 15-10). This will ensure bark will not be stripped from the trunk when the limb is separated. Correct cutting of large limbs will leave only a small heel of the former branch. Long stubs are unattractive (Figure 15-11). When pruning smaller branches, it is best to make a cut just above a node, the place where leaves and axillary buds emerge from the stem (Figure 15-12). It is a good idea to stand back from the plant to get an idea of the general vicinity of where to make the cut on the stem, striving to maintain good geometric form.

Pruning needs to be accomplished when light levels and duration are low and plants exhibit signs of **etiolation**, where growth tips search for light and the plant compensates with long **internodes** and small, pale leaves. Pruning or, in the case of smaller plants or vining plants like *Epipremnum* or *Tradescantia*, pinching is needed to keep the plant compact.

The pruning of strong leader-type plants like *Dracaena marginata* is much more dramatic. If the foliage portion is pruned because a plant is growing too tall, all that will be left is a leafless stem. Indoors, the plant could remain that way for months only to sport a few, immature leaves after that time. This does not provide the necessary beauty associated with healthy indoor plants. In situations such as these, it is best to remove large, but healthy, plants and place them in sites that can hold larger, taller specimens well, or sell them outright.

FIGURE 15-11: Stub left on this tree (a) is unattractive while (b) is correct.

(a)

(b)

SUMMARY

Interior plantscaping requires practice in order to master its many divisions including sales, cleaning, display, and design. It is an active area of science, design, and business, so in order to be good at it, practice improves the horticulturist's chance for success.

There are many challenges that face an interiorscaper. Large installations or management and ownership provide daily situations where success is possible, but not certain. Displaying plants in refreshing, innovative ways brings more attention to the pleasure plants provide for people. In order to keep plants looking good, it is necessary to master many techniques in display and maintenance of indoor plants.

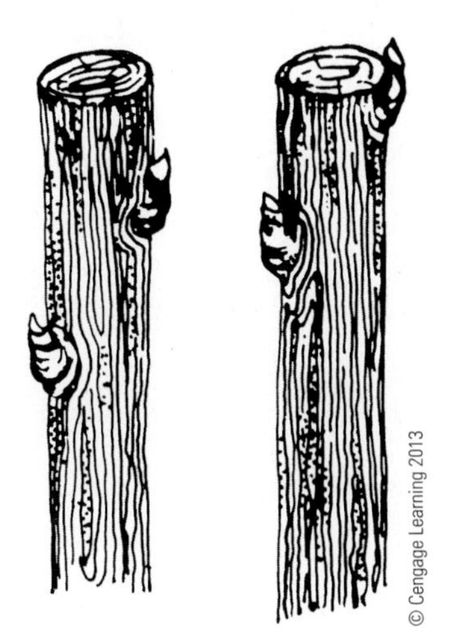

FIGURE 15-12: Prune branches just above a bud. On the left, the cut is too far above the bud while the one on the right is too close.

APPENDIX

INTERIOR PLANT LIST

Pronunciation

Names of plants in Latin should use Classical Latin pronunciation while names adapted from people or places should follow their original pronunciation, followed by Latin endings. The following table is a guide for Classical Latin pronunciation. Try these sounds on the following plant materials list.

Latin Letter or Diphthong	**Classical Pronunciation**
a	Short *a* is pronounced as either in h*a*t or in *a*llow. Long *a* as in g*a*ther, never as in *a*ce.
ae	As *i* in k*i*te.
au	As *ou* in *ou*ch.
c	Always hard as in *c*olor.
e	Short *e* as in g*e*t, long *e* as *a* in d*a*te.
ei	As *a* in l*a*te.
eu	As *eu* in f*eu*d.
g	Always hard as in *g*et.
i	Short *i* as in k*i*n, long *i* as *ee* in s*ee*n.
j	As *y* in *y*ellow.
o	Short as in r*o*t, long as in v*o*te.
oe	As *oi* in *oi*l.
ph	As *p* or *p-h*.
r	Always trilled or rolled.
s	As in thi*s*.
t	As in *t*oggle.
u	Short *u* as in b*u*ll; long *u* as *oo* in b*oo*t, never as in r*u*b.
ui	as in d*ewy*.
v	Classically pronounced as *w*. As in *w*ork.
y	From Greek origins, short as in f*oo*t, long as in *u*ber

Writing scientific names requires precision, but their pronunciation is not governed by rules. Most people pronounce the names of plants to suit their own language. It is not unusual to hear plant names pronounced in different ways, varying geographically, institutionally, or disciplinarily.

Plants

Adiantum raddianum

FIGURE A-1: *Adiantum raddianum*, Maidenhair Fern, Delta Maidenhair Fern

Pronunciation: a-dee-AHN-tum roddy-AH-num

Common name: Maidenhair Fern, Delta Maidenhair Fern

Family name: *Pteridaceae*

Description: Evergreen fern with short rhizomes and somewhat triangular, 3- or 4-pinnate, black-stemmed fronds, to 24 inches long (60 centimeters) with irregularly triangular lobed segments. Tropical North and South America, West Indies.

Maintenance: Bright, indirect light and constantly moist, but not wet, soil. Well-suited for terrarium culture.

Propagation: Spores, division of rhizomes in early spring.

Pests/Disease: Scale.

Notes: The native habitats of these ferns are near water sources or in shady crevices. *Adiantos* is Greek, meaning unwetted, in reference to the way the foliage repels water. Rounded sori form on the frond margins (Figure A-1).

Adonidia merrillii adonidia palm

Pronunciation: ah-don-IH-dee-uh MER-il-ee-ee

Common name: Manila Palm, Christmas Palm

Family name: *Arecaceae*

Description: Solitary stems tan in color, crowned by a tuft of deep green fronds; petioles long and clasping provide distinctive light green color between fronds and trunk.

Maintenance: Bright, filtered light.

Propagation: Seed.

Pests/Disease: Fungal leaf spots.

Notes: In the exterior, the plant produces bright red fruit in winter; hence, common name Christmas Palm.

Aechmea fasciata

Pronunciation: ike-MEE-uh fass-kee-AH-tuh

Common name: Silver Vase Plant, Vase Plant

Family name: *Bromeliaceae*

Description: Rhizomatous, epiphytic perennial with funnel-shaped rosettes of silvery-gray and green leaves, the silver color is caused by minute scales; spines occur on leaf margins; inflorescences have blue petals, rose-pink bracts, and sepals.

Maintenance: Bright light in loose, well-drained potting mix, keep foliage vase filled with water.

Propagation: Pot offsets taking care to not place them too deep in soil to avoid rot.

Pests/Disease: Scale, mealybugs, leaf spot, crown rot.

Notes: Native habitat is the rainforest in Central and South America, most *Aechmea* are epiphytic (Figure A-2).

FIGURE A-2: *Aechmea fasciata*, Silver Vase Plant, Vase Plant

Aeonium sp.

Pronunciation: eye-O-nee-um

Common name: Aeonium

Family name: *Crassulaceae*

Description: Succulent plants, often sub-shrublike, with rosette leaves atop clustered shoots.

Maintenance: Require a well-drained soil; allow drying between watering, high light levels indoors.

Propagation: Allow rosette cuttings to callus, then pot in well-drained, sandy mix.

Pests/Disease: Aphids, mealybugs, fungal rots, and spots.

Notes: Native to North Africa, Canary Islands, and the Mediterranean (Figure A-3).

FIGURE A-3: *Aeonium sp.*, Aeonium

Aeschynanthus lobbianus

Pronunciation: ice-skee-NAN-thus lobby-AN-us

Common name: Lipstick Plant

Family name: *Gesneriaceae*

Description: Ovate to lance-shaped leaves in opposite pairs or whorls; flowers produced in pairs in leaf axils near stem tips in summer to winter, prominent calyces and often tubular corollas when in bud give appearance of lipstick in a tube.

Maintenance: Bright filtered light, high humidity.

Propagation: Stem cuttings, seed.

Pests/Disease: Aphids, thrips, scale, mealybugs, fungal leaf spots, *Rhizoctonia* blight, Tobacco mosaic virus.

Notes: From Greek *aischyno* (shame) and *anthos* (a flower) (Figure A-4).

FIGURE A-4: *Aeschynanthus lobbianus*, Lipstick Plant

Agave sp.

FIGURE A-5: *Agave horrida*, Agave

Pronunciation: a-GAH-vay
Common name: Agave
Family name: *Agavaceae*
Description: Rosette-forming
Maintenance: Cactus potting mix in full light; keep drier in winter.
Propagation: Pot offsets in sharp sand and peat if unrooted; if rooted, treat as mature plant, seed.
Pests/Disease: Scale, mealybugs, fungal blights, root rot, bacterial soft rot.
Notes: From Greek *agave* (noble), American desert and mountain regions.

Agave horrida (A. HOR-ri-da) has sharp marginal spines (Figure A-5).

Aglaonema commutatum

FIGURE A-6: *Aglaonema commutatum*, Chinese Evergreen

Pronunciation: a-glah-oh-NEE-ma com-mu-TAY-tum
Common name: Chinese Evergreen
Family name: *Araceae*
Description: Usually rhizomatous, erect or recumbent short stems; sporadic, small spathe and spadix flowers.
Maintenance: Filtered light, high humidity, allow to somewhat dry out between waterings to avoid stem or root rot.
Propagation: Stem cuttings with three to four leaves in light media or water, division.
Pests/Disease: Mealybugs, scale, bacterial and fungal rots caused by overwatering.
Notes: Native to tropical Asia (Figure A-6).

Aglaonema modestum (mo-DES-tum) Chinese Evergreen has dark green leaf blades and is an erect-growing perennial native to southern China and Thailand.

Alocasia sanderiana

Pronunciation: a-low-KAY-see-a san-dare-ee-AH-na
Common name: Kris Plant
Family name: *Araceae*
Description: Rhizomatous perennial with arrow-shaped leaves with purplish-green ground and prominent green veins.

Maintenance: Filtered light, high humidity.

Propagation: Divide rhizomes or pot offsets, seed propagation.

Pests/Disease: Mealybugs, scale, fungal and bacterial leaf diseases.

Notes: Native to the Philippines (Figure A-7).

© Cengage Learning 2013

FIGURE A-7: *Alocasia sanderiana*, Kris Plant

Aloe vera

Pronunciation: AL-o VAIR-a

Common name: Aloe

Family name: *Aloeaceae*

Description: Rosette-leaved perennial with lance-shaped, succulent leaves; racemes of flowers in greenhouse culture, but grown for attractive foliage indoors.

Maintenance: Bright light, well-drained soil containing sand, avoid over-watering.

Propagation: Pot offsets or propagate by seed.

Pests/Disease: Mealybugs, scale, root rot, soft rot, fungal stem, and leaf rots.

Notes: Native to tropical and southern Africa, Madagascar, and the Arabian Peninsula. Gel from leaf sap used for burn treatment (Figure A-8).

© Cengage Learning 2013

FIGURE A-8: *Aloe vera*, Aloe

Ananas comosus

Pronunciation: a-NAN-nas cuh-MO-sus

Common name: Pineapple

Family name: *Bromeliaceae*

Description: Evergreen bromeliads.

Maintenance: Bright light, freely draining soil, keep barely moist.

Propagation: Root offsets; remove leafy rosette from top of fruit, allow to callus, and then root in peat/sand mix.

Pests/Disease: Mealybugs, scale, fungal root rot.

Notes: Native to South America, produce flowers in a dense terminal inflorescence. Sharp-toothed leaf margins (Figure A-9).

© Cengage Learning 2013

FIGURE A-9: *Ananas comosus*, Pineapple

Anthurium andraeanum

Pronunciation: an-THEWR-ree-um an-dree-AH-num

Common name: Anthurium

Family name: *Araceae*

FIGURE A-10: *Anthurium andraeanum*, Anthurium

Description: Large, entire leaves, flowering with brightly colored spathes and cylindrical spadices in red, white, pink, orange, and more.

Maintenance: Keep crowns just above soil line to prevent rot, maintain in filtered light and high humidity.

Propagation: Root stem cuttings, divide offsets or rootstock, sow seed.

Pests/Disease: Mealybugs, scale, fungal root rot, *Xanthomonas*, fungal leaf spots, bacterial soft rot.

Notes: From mountain forests of tropical and subtropical North and South America (Figure A-10).

Anthurium scherzerianum (sure-zur-ee-AH-num) Pigtail Plant has curling spadices.

Aphelandra squarrosa

Pronunciation: af-el-AN-dra sqwar-ro-sa

Common name: Zebra Plant

Family name: *Acanthaceae*

Description: Leaves opposite with white, yellow, or silver veins, attractive flower heads of bright-colored yellow bracts and flowers.

Maintenance: Bright, indirect light.

Propagation: Remove flower spike when spent to encourage side-shoot production, and then root cuttings.

Pests/Disease: Aphids, mealybugs, fungal disease susceptibility.

Notes: From moist woodland habitats in tropical North, South, and Central America. Older plants deteriorate rapidly and should be propagated or composted.

FIGURE A-11: *Araucaria heterophylla*, Norfolk Island Pine

Araucaria heterophylla

Pronunciation: a-row-KAY-ree-a he-te-ro-FILL-a

Common name: Norfolk Island Pine

Family name: *Araucariaceae*

Description: Fan-like, wedge-shaped, lacy foliage.

Maintenance: Bright, indirect light.

Propagation: Vertical shoot tip cuttings, seed.

Pests/Disease: Mealybugs, scale.

Notes: Native to Norfolk Island, between New Caledonia, New Zealand, and Australia (Figure A-11).

Ardisia crispa

Pronunciation: ar-DISS-e-a KRIS-pa
Common name: Coral Berry
Family name: *Myrsinaceae*
Description: Spiraled or alternate, glossy, scalloped-margined leaves, corymb inflorescences of pink flowers produce bright red berries in summer.
Maintenance: Bright, filtered light; light pruning when necessary for symmetry.
Propagation: Cuttings, seed.
Pests/Disease: Mealybugs, fungal root rots.
Notes: Native to Japan and southern China (Figure A-12).

FIGURE A-12: *Ardisia crispa*, Coral Berry

Asparagus densiflorus 'sprengeri'

Pronunciation: a-SPA-ra-gus dens-i-FLOR-us spreng-a-REE
Common name: Sprengeri, Sprenger's Asparagus
Family name: *Liliaceae*
Description: Evergreen tuberous perennial, arching stem-like leaves emerge with attached, leaf-like cladophylls providing an overall feathery effect; small white flowers followed by red berries.
Maintenance: Bright, filtered light; lower light levels or dry soil cause cladophyll drop.
Propagation: Division, seed.
Pests/Disease: Spider mites, aphids, crown rot.
Notes: Widely used in floral design, the stems may bear spines (Figure A-13).

Asparagus densiflorus 'Myers', Foxtail Fern, dense, arching, foxtail-like fronds (Figure A-14).

Asparagus macowanii (A. ma-COW-an-ee-ee), Ming Fern, upright stems resembling bonsai pine (Figure A-15).

Asparagus setaceus (A. say-TAH-kee-us), (=*A. plumosus*) Plumosa fern, fine textured, sheer to feathery, dark green leaves (Figure A-16).

FIGURE A-13: *Asparagus densiflorus 'Sprengeri'*, Sprengeri, Sprenger's Asparagus

Aspidistra elatior

Pronunciation: a-spi-DI-stra ay-LAY-tee-or
Common name: Aspidistra, Cast Iron Plant
Family name: *Liliaceae*
Description: Rhizomatous perennial, lance-shaped, dark green leaves.

FIGURE A-14: *Asparagus densiflorus 'Myers'*, Foxtail Fern

FIGURE A-15: *Asparagus macowanii*, Ming Fern

FIGURE A-16: *Asparagus setaceus*, Plumosa fern

FIGURE A-17: *Aspidistra elatior*, Aspidistra, Cast Iron Plant

Maintenance: Bright, filtered light, but tolerant of low-light levels.

Propagation: Division.

Pests: Mealybugs, scale, spider mites, fungal leaf spots.

Notes: Maroon flowers, 1 inch (2 to 3 centimeters) borne along rhizomes at soil surface of mature plants. Foliage harvested for floral designs (Figure A-17).

Asplenium nidus

Pronunciation: a-SPLAY-nee-um NEE-dus

Common name: Bird's Nest Fern

Family name: *Aspleniaceae*

Description: Fiber-covered rhizome from which ovate to lance-shaped, bright green leaves emerge, forming rosette in which organic material collects in the wild.

Maintenance: Bright, filtered light and moderate humidity.

Propagation: Repot plantlets or sow spores.

Pests/Disease: Scale, mealybugs, bacterial leaf diseases.

Notes: *Splen*, from the Greek, meaning spleen, *nidus*, nest (Figure A-18).

Beaucarnea recurvata (syn. with nolina recurvata)

Pronunciation: bow-CAR-nay-a re-kur-WA-ta

Common name: Ponytail Palm

Family name: *Agavaceae*

Description: Long, recurving leaves borne in a terminal rosette atop a flask-shaped stem, older plants may branch.

Maintenance: Bright light taking care not to overwater.

Propagation: Root offsets or sow seed.

Pests: Spider mites, scale, fungal spots, stem rots, bacterial soft rot.

Notes: Found in Southeast Mexico (Figure A-19).

Begonia rex-cultorum

Pronunciation: bay-GON-ee-a reks-kul-TO-rum

Common name: Rex Begonia

Family name: *Begoniaceae*

Description: Usually rhizomatous, grown for their brilliantly colored, highly patterned leaves.

Maintenance: Bright, indirect light, floral removal keeps plant's energy in foliage production; watering best accomplished by immersing pots, and then allowing to thoroughly drain.

Propagation: Rhizome or leaf cuttings, sow seed.

Pests/Disease: Mealybugs, mites, thrips, whiteflies, powdery mildew, stem rot, rhizome rot, nematodes.

Notes: Named after Michael Begon (1638–1710), Governor of French Canada; these plants were prized in Victorian households. *Rex* is Latin for king (Figure A-20).

The American Horticultural Society divides begonias into seven informal groupings based on growth habits and maintenance needs. In addition to *Rex-cultorum* begonias, there are:

Cane-stemmed begonias with bamboo-like stems, asymmetrical and often deeply lobed leaves and showy flowers; Rhizomatous begonias are grown for their flowers and foliage; *Semperflorens* begonias, *Begonia x semperflorens,* Wax Begonias, are widely produced for their bushy habit and constant bloom. *Semperflorens* are often used in outdoor flower beds and containers. Shrub-like begonias can bloom seasonally, occasionally, or are ever blooming and stems may be branching, erect, or semi-erect. Tuberous begonias are winter-dormant and are grown for their showy flowers. Dormant tubers should be planted with the hollow side up. Winter-flowering begonias (Elatior [*B. X hiemalis*] including Rieger hybrids) (Figure A-21) are widely produced as potted plants, are usually fibrous-rooted but sometimes have tuber-like roots, and flower in bright, indirect light.

FIGURE A-18: *Asplenium nidus*, Bird's Nest Fern

FIGURE A-19: *Beaucarnea recurvata* (Syn. with *Nolina recurvata*), Ponytail Palm

Bougainvillea glabra

Pronunciation: boo-gan-VIL-lee-a GLA-bra

Common name: Bougainvillea

Family name: *Nyctaginaceae*

Description: White, tubular flowers surrounded by three colorful bracts in bright pink, orange, lavender, and others; stems may be thorny, alternate leaves are oval with pointed tips.

Maintenance: Bright light, balanced fertilizer to encourage blooming.

Propagation: Root cuttings.

Pests/Disease: Spider mites, whiteflies, aphids, bacterial and fungal leaf spots.

Notes: Named for Louis Antoine de Bougainville (1729–1811), scientist and explorer. Native to tropical and subtropical South America, these plants are well suited to porch and balcony plantings.

FIGURE A-20: *Begonia Rex-cultorum*, Rex Begonia

FIGURE A-21: *Begonia X hiemalis*, Winter-flowering Begonia

Brugmansia sp.

Pronunciation: brug-MAN-see-a
Common name: Angels' Trumpets
Family name: *Solanaceae*
Description: Large, pendant, trumpet-shaped flowers in whites, yellows, pinks, and oranges; leaves alternate.
Maintenance: Bright light.
Propagation: Vegetative cuttings or sow seed.
Pests/Disease: Spider mites, whiteflies, mealybugs.
Notes: Formerly included in genus *Datura*, *Brugmansia* differ having pendulous flowers rather than erect. Native to southern United States to South America.

FIGURE A-22: *Bucida buceras*, Black Olive

Bucida buceras

Pronunciation: bew-SYE-duh bew-SER-azz
Common name: Black Olive
Family name: *Combretaceae*
Description: Large evergreen tree with smooth bark and somewhat horizontal, spreading habit when mature; alternate leaves, oblanceolate to obovate; drooping, graceful branches.
Maintenance: Provide bright light indoors, use well-drained soil.
Propagation: Air layering, seed.
Pests/Disease: Mealybugs, scale.
Notes: Found in South Florida landscapes. Plant provides highly patterned effect indoors (Figure A-22).

FIGURE A-23: *Calathea sp.*, Calathea

Calathea sp.

Pronunciation: ka-LAH-thee-a
Common name: Calathea
Family name: *Marantaceae*
Description: Ovate to elliptic leaves, highly patterned above and often maroon on undersides, clump forming.
Maintenance: Bright, indirect light with high humidity, not tolerant of drafts.
Propagation: Division.
Pests/Disease: Spider mites, mealybugs, aphids, fluoride toxicity, marginal necrosis due to low humidity, fungal and bacterial leaf spots.

Notes: From the Greek *kalathos* (a basket), the flowers are borne within sheathing bracts (Figure A-23).

Calceolaria herbiohybrida group

Pronunciation: kal-kee-oh-LAH-ree-a
Common name: Pouch Flower, Slipper Flower
Family name: *Scrophulariaceae*
Description: Compact, biennials with opposite, softly hairy leaves, dense flower clusters in red, yellow, orange, and bicolor.
Maintenance: Keeping plants at cool indoor temperatures prolongs bloom duration. As flowers decline, the plant is no longer usable for display and should be replaced.
Propagation: Seed.
Pests/Disease: Spider mites, whiteflies, aphids, *Botrytis*.
Notes: Grown as potted plants for short-term color. From the Latin *scrofulae*, a swelling of the neck or glands.

Carissa grandiflora

Pronunciation: ka-RIS-sa gran-di-FLOR-a
Common name: Natal Plum
Family name: *Apocynaceae*
Description: Small plant to shrub; oval, dark green leaves, fragrant white flowers.
Maintenance: Bright, filtered light.
Propagation: Stem cuttings, seed.
Pests: Mealybugs, scale, Anthracnose.
Notes: Native to South Africa.

Caryota mitis

Pronunciation: ka-ree-OH-ta MEE-tis
Common name: Fishtail Palm
Family name: *Arecaceae*
Description: Asymmetrical pinnae have appearance of fish fins, juvenile stems covered in fine brown fibers; to 20 feet (7 meters) indoors.
Maintenance: Bright, filtered light; high humidity.
Propagation: Seed.
Pests/Disease: Spider mites, scale, fungal leaf spots.
Notes: Burma, Malaysia, Indonesia, Philippines are native habitats (Figure A-24).

FIGURE A-24: *Caryota mitis*, Fishtail Palm

FIGURE A-25: *Cattleya sp.*, Cattleya Orchid

FIGURE A-26: *Ceropegia woodii*, Rosary Vine, String of Hearts

FIGURE A-27: *Chamaedorea elegans*, Bamboo Palm

Cattleya sp.

Pronunciation: CAT-lee-a

Common name: Cattleya Orchid

Family name: *Orchidaceae*

Description: Distinguished from other orchids by club-shaped pseudobulbs on short rhizomes from which emerge strap-shaped, sturdy green leaves. Terminal, showy flowers, most often with a prominent lip, emerge from bract-like leaves singly or in racemes.

Maintenance: Using epiphytic orchid mix, grow the plants in orchid baskets or pots in bright light, high humidity.

Propagation: Division, tissue culture.

Pests/Disease: Scale, spider mites, whiteflies, aphids, mealybugs, fungal and bacterial leaf spots can affect pseudobulbs, leaf rots, viruses.

Notes: Named for William Cattley, nineteenth-century horticulturist. Many cultivars exist in white, lavender, orange, green, bicolor, and more (Figure A-25).

Ceropegia woodii

Pronunciation: kay-row-PEE-gee-a WUD-ee-ee

Common name: Rosary Vine, String of Hearts

Family name: *Asclepiadiaceae*

Description: Tuberous-rooted succulent with fine stems up to 3 feet (1 meter) long, leaves are heart-shaped, silver and green with purple undersides, with bulbils produced in leaf axils.

Maintenance: Display as a hanging plant and provide with bright, indirect light; do not overwater and use a well-drained potting mix.

Propagation: Pot bulbils, stem cuttings with nodes or seed.

Pests/Disease: Aphids, scale, mealybugs.

Notes: Native to South Africa (Figure A-26).

Chamaedorea elegans

Pronunciation: ka-mie-DOE-ree-a AY-le-gans

Common name: Bamboo Palm

Family name: *Palmae*

Description: Compact palm of deep green leaves, height to about 24 inches (60 centimeters) indoors.

Maintenance: Bright, indirect light; keep from direct sunlight, which will encourage spider mite infestation.

Propagation: Division or sow seeds.

Pests/Disease: Spider mites, scale, mealybugs, fungal leaf spots, root rot.

Notes: Also known as *Neanthe bella* Parlor Palm (Figure A-27).

Chamaerops humilis

Pronunciation: ka-MEE-rops HUM-i-lis

Common name: Dwarf Fan Palm

Family name: *Palmae*

Description: Low growing, to 36 inches (90 centimeters), broad, palm-shaped leaves.

Maintenance: Bright to indirectly bright light, well-drained soil.

Propagation: Separate sucker plants and pot separately or sow seed.

Pests/Disease: Spider mites.

Notes: Native to the Mediterranean region.

Chlorophytum comosum

Pronunciation: klo-ROF-i-tum ko-MOE-sum

Common name: Spider Plant

Family name: *Liliaceae*

Description: Perennial with linear leaves and arching racemes producing small, white flowers and plantlets; fleshy roots.

Maintenance: Bright, indirect light.

Propagation: Pot plantlets, divide mature plants.

Pests/Disease: Scale, leaf tip necrosis due to soluble salt accumulation or overly dry soil.

Notes: *Comosum* (tufted). Well-suited to displays where racemes can freely cascade such as in wall planters or hanging baskets.

Chlorophytum comosum 'Variegata' (va-ri-GAH-ta), Variegated Spider Plant (Figure A-28).

FIGURE A-28: *Chlorophytum comosum* 'Variegata', Variegated Spider Plant

Chrysalidocarpus lutescens

Pronunciation: kri-sah-li-doe-KAR-pus loo-TES-ens

Common name: Areca Palm

Family name: *Palmae*

Description: Pinnate leaves 3 to 6 feet (1 to 2 meters) long, stems grow in clusters.

Maintenance: Bright, indirect light, water when soil is dry to the touch. Avoid dry areas or increase humidity.

FIGURE A-29: *Chrysalidocarpus lutescens*, Areca Palm

Propagation: Division, seed.

Pests: Spider mites, whiteflies, scale, mealybugs, fungal leaf spots.

Notes: *Lutescens* is Latin for yellow, in this case, the yellow-green coloring of the stems and leaves (Figure A-29).

Cissus rhombifolia

Pronunciation: KIS-us rom-bi-FO-li-a

Common name: Grape Ivy

Family name: *Vitaceae*

Description: Tri-leaflet leaves, dark green with soft hairs; vigorous climber or cascading if without support.

Maintenance: Bright, indirect light, avoid water splash on leaves, which promotes mildew. Growing plants may require pinching to avoid legginess.

Propagation: Vegetative cuttings.

Pests/Disease: Whiteflies, spider mites, mealybugs, powdery mildew, stem and root rot.

Notes: This plant is in the grape family.

Clivia miniata

Pronunciation: KLY-vee-a min-ee-AH-ta

Common name: Kafir Lily

Family name: *Amaryllidaceae*

Description: Strap-shaped leaves, dark green, up to 24 inches (60 centimeters) long, funnel-shaped florets in umbels of orange, yellow, or red.

Maintenance: Bright filtered, indirect light; restricted root environment encourages blooming.

Propagation: Division, sow seed.

Pests/Disease: Mealybugs, bacterial and fungal spots.

Notes: Native to South Africa. Genus named after Lady Charlotte Florentina Clive, Duchess of Northumberland, and daughter of Robert Clive who was instrumental in securing India for Britain in the eighteenth century.

Codiaeum variegatum

Pronunciation: co-dee-*I*-um va-ree-a-GAH-tum

Common name: Croton

Family name: *Euphorbiaceae*

Description: Grown for colorful patterned foliage, with many cultivars and leaf forms, small tree or shrub; smallish flowers produced in axillary racemes are usually removed to keep plant's energy in foliage maintenance.

Maintenance: Bright light, water when soil is dry to the touch; drafts and cold temperatures may cause leaf drop.

Propagation: Cuttings, air layering.

Pests/Disease: Spider mites, scale, mealybugs, root rot, fungal and bacterial leaf spots.

Notes: Contact with foliage may irritate skin (Figure A-30).

FIGURE A-30: *Codiaeum variegatum*, Croton

Cordyline terminalis

Pronunciation: kor-DI-li-nee ter-min-AH-lis

Common name: Ti Plant

Family name: *Agavaceae*

Description: Clump-forming, small shrub, strap-shaped leaves, deep greenish-purple with magenta edges.

Maintenance: Bright filtered light, well-drained soil.

Propagation: Vegetative cuttings, sow seed.

Pests/Disease: Spider mites, scale, mealybugs, bacterial and fungal spots, fluoride toxicity producing necrotic leaf tips and margins.

Notes: Originates in tropical Southeast Asia, Australia, and the Pacific Islands (Figure A-31).

FIGURE A-31: *Cordyline terminalis*, Ti Plant

Crassula ovata

Pronunciation: KRAS-ew-la oh-VAY-ta

Common name: Jade Plant

Family name: *Crassulaceae*

Description: Branched, succulent plant to small shrub; thick, fleshy stems and leaves, leaves oval in shape, often with red margins; white, star-shaped flowers.

Maintenance: Cactus potting mix, bright light, water somewhat sparingly, avoid cold drafts and overwatering.

Propagation: Root stem or leaf cuttings, sow seed.

Pests/Disease: Mealybugs, aphids, *Botrytis*, fungal leaf spots, stem and root rot.

Notes: From the Latin *crassus* (thick) and *ovata* (egg-shaped) in reference to the leaves. Native to South Africa (Figure A-32).

FIGURE A-32: *Crassula ovata*, Jade Plant

Crossandra infundibuliformis

Pronunciation: kros-AN-dra in-fun-di-bew-lee-FORM-is
Common name: Firecracker Flower, Crossandra
Family name: *Acanthaceae*
Description: Glossy, deep green leaves, fan-shaped orange flowers.
Maintenance: Bright light, out of direct sun which may burn leaves, pinch leggy growth.
Propagation: Vegetative cuttings, seed.
Pests/Disease: Aphids, thrips, whiteflies, stem and root rot.
Notes: Native to southern India, Sri Lanka.

Cryptanthus bivittatus

FIGURE A-33: *Cryptanthus bivittatus*, Earth Star

Pronunciation: krip-TANTH-us bi-vi-TAH-tus
Common name: Earth Star
Family name: *Bromeliaceae*
Description: Terrestrial bromeliads, stemless, offsets form in leaf axils, leaves to 4 inches (10 centimeters) long, cultivated varieties in varying colors.
Maintenance: Bright to indirect light, high humidity, potting mix high in organic matter; well-suited for terrarium culture.
Propagation: Pot offsets, sow seed.
Pests/Disease: Scale, mealybugs.
Notes: From the Greek *krypto* (to hide) and *anthos* (a flower). Native to eastern Brazil (Figure A-33).

Ctenanthe sp.

FIGURE A-34: *Ctenanthe sp.*, Ctenanthe

Pronunciation: ten-ANTH-ee
Common name: Ctenanthe
Family name: *Marantaceae*
Description: Patterned leaves with maroon undersides, leaves ovate to linear depending on species, height of species varying from 12 to 36 inches (30 to 90 centimeters).
Maintenance: Bright, filtered light with high humidity, soil mix high in organic material. Do not allow soil to dry.
Propagation: Division or seed.
Pests: Mealybugs, spider mites.
Notes: Native to Brazilian and Costa Rican forest floors. Named from the Greek *kteinos* (a comb) and *anthos* (a flower), in reference to the comb-shaped flower bracts (Figure A-34).

Cycas revoluta

FIGURE A-35: *Cycas revoluta,* Japanese Sago Palm

Pronunciation: SY-kas re-vo-LOO-ta
Common name: Japanese Sago Palm
Family name: *Cycadaceae*
Description: Upright growth at juvenile stages, then spreading and suckering with age, arching leaves, 30 to 60 inches (.75 to 1.5 meters) long with stiff, needle-like, dark green leaflets.
Maintenance: Bright light, allow soil to dry to the touch prior to watering.
Propagation: Pot suckering plants, sow seeds.
Pests/Disease: Mealybugs, scale, spider mites, leaf spots, root rot.
Notes: Native to Ryuku Islands, Japan (Figure A-35).

Cyclamen persicum

FIGURE A-36: *Cyclamen persicum*, Florists' Cyclamen

Pronunciation: SIKE-la-men PER-si-kum
Common name: Florists' Cyclamen
Family name: *Primulaceae*
Description: Tuberous perennial with heart-shaped leaves having silvery patterns above and rosy-purple undersides, flowers borne above foliage crown in pinks, whites, and purples.
Maintenance: Provide bright, filtered light; keeping temperatures cooler will result in longer bloom, and hence display duration.
Propagation: Plant tuber tops just above soil surface and grow in bright, filtered light or sow seed in cool greenhouse.
Pests/Disease: Spider mites, cyclamen mites, *Botrytis.*
Notes: Good for winter displays indoors and can tolerate cool temperatures in window boxes and protected outdoor displays (Figure A-36).

Cymbidium sp.

Pronunciation: kim-BID-ee-um
Common name: Cymbidium Orchid
Family name: *Orchidaceae*
Description: A cool temperature greenhouse orchid; long, strap-like leaves emerging from rounded pseudobulbs, flowers borne on long racemes in the spring, sometimes with as many as 50 flowers.
Maintenance: Orchid mix or potting soil mixed with bark, bright light; water thoroughly and allow to drain.

© Cengage Learning 2013

FIGURE A-37: *Cymbidium sp.*, Cymbidium Orchid inflorescence

© Cengage Learning 2013

FIGURE A-38: *Cyperus alternifolius*, Umbrella Plant

© Cengage Learning 2013

FIGURE A-39: *Davallia fejeensis*, Rabbit's Foot Fern

Propagation: Division, tissue culture.

Pests/Disease: Spider mites, scale, mealybugs, root rot, *Botrytis*, bacterial soft rot.

Notes: Widely used as a cut flower and potted plant; many cultivated varieties exist in yellow, white, mauve, and other colors (Figure A-37).

Cyperus alternifolius

Pronunciation: si-PE-rus al-ter-ni-FO-li-us

Common name: Umbrella Plant

Family name: *Cyperaceae*

Description: Long, dark green stems produce a terminal tuft of leaf-like bracts below compound umbels of tan or brown flowers, which are produced in the summer.

Maintenance: Grow plants with pots standing in saucers of water or completely submerged for this moisture-loving plant.

Propagation: Float leafy inflorescences in water, roots will emerge from axils; divide clumps or sow seed.

Pests/Disease: Mealybugs.

Notes: Native to wet habitats (Figure A-38).

Davallia fejeensis

Pronunciation: da-VALL-ee-a fee-jee-EN-sis

Common name: Rabbit's Foot Fern

Family name: *Davalliaceae*

Description: Broad green fronds produced from tan, scale-covered, thick rhizomes.

Maintenance: Bright, indirect light in a potting mix high in organic matter kept evenly moist but not soggy.

Propagation: Division of root-bearing rhizomes or sow spores.

Pests/Disease: Scale, *Botrytis*.

Notes: Native to Fiji, suitable for hanging baskets where the creeping rhizomes may be seen (Figure A-39).

Dendranthema grandiflora

Pronunciation: den-dran-THEEM-a gran-di-FLOW-rum

Common name: Chrysanthemum

Family name: *Asteraceae*

Description: Deep green leaves surmounted by composite flowers in forms such as daisy, decorative, button, spider, and spoon; color range in white, pink, purple, bronze, yellow, and green.

Maintenance: Bright, indirect light; keep plants cool for prolonged display, water when soil is dry to the touch.

Propagation: Vegetative cuttings, sow seeds, division of large plants.

Pests/Disease: Aphids, spider mites, whiteflies, leaf miners, fungal rot, *Botrytis*, powdery mildew.

Notes: Remove shipping sleeves and keep in florist cooler for up to two weeks to delay senescence.

Dendrobium sp.

Pronunciation: den-DROE-bee-um

Common name: Dendrobium Orchid

Family name: *Orchidaceae*

Description: Epiphytic and terrestrial orchids from lowland rainforests to high mountain forests, long pseudobulbs, lance-shaped to oval leaves, showy flowers on racemes or panicles.

Maintenance: Epiphytic orchid mix, bark slabs or orchid baskets allow root aeration, filtered light, high humidity; flowering often more prolific when confined to smaller containers, inflorescences may need support.

Propagation: Division, pot offsets, stem cuttings, tissue culture.

Pests/Disease: Spider mites, aphids, whiteflies, leaf spots, *Botrytis*, bacterial and fungal rots.

Notes: Many cultivated varieties exist in lavender, white, magenta, green, bicolor, and others (Figure A-40).

FIGURE A-40: *Dendrobium sp.*, Dendrobium Orchid

Dieffenbachia seguine (syn. with d. maculata and d. picta)

Pronunciation: dee-fan-BAHK-ee-a se-GWEE-nay

Common name: Dumb Cane

Family name: *Araceae*

Description: Alternate oval to lance-shaped, broad leaves, 12 to 18 inches (30 to 45 centimeters) long with various patterns, often white on green, plants generally 3 to 10 feet (1 to 3 meters) tall.

Maintenance: Bright, filtered light, high humidity; water when soil is dry to the touch. Overwatering and cold temperatures cause disease susceptibility.

Propagation: Stem cuttings or stem segment cuttings, air layering.

Pests/Disease: Mealybugs, scale, spider mites, aphids, stem and root rot, bacterial blight, soft rot.

Notes: Named for J. F. Dieffenbach (1790–1863), head gardener of the royal palace in Vienna circa 1830. Synonymous with *D. amoena* and *D. picta* (Figure A-41).

FIGURE A-41: *Dieffenbachia seguine*, Dumb Cane

FIGURE A-42: *Dracaena deremensis 'Janet Craig'*, Janet Craig Dracaena

Dracaena deremensis 'janet craig'

Pronunciation: dra-KEE-na de-rem-EN-sis
Common name: Janet Craig Dracaena
Family name: *Dracaenaceae*
Description: Height to 6 feet (2 meters), dark green, strap-like leaves with longitudinal ripples.
Maintenance: Bright, indirect light, tolerant of lower light levels, water when soil is dry to the touch.
Propagation: Stem tip cuttings, stem segment cuttings, air layering.
Pests/Disease: Mealybugs, spider mites, scale, fluoride toxicity, boron and calcium deficiencies.
Notes: Hardy indoor plant, wipe dusty leaves with a soft cloth to reveal natural gloss (Figure A-42).

FIGURE A-43: *Dracaena deremensis 'Warnecki'*, Warnecki Dracaena

Dracaena deremensis 'warnecki'

Pronunciation: dra-KEE-na de-rem-EN-sis WAR-nek-ee
Common name: Warnecki Dracaena
Family name: *Dracaenaceae*
Description: Strap-like, gray-green leaves with white longitudinal stripes, 16 to 24 inches (40 to 60 centimeters) long, upright growth habit.
Maintenance: Bright, indirect light, tolerant of lower light levels, water when soil is dry to the touch; wipe dusty leaves with a soft cloth to reveal natural gloss.
Propagation: Stem tip cuttings, stem segment cuttings, air layering.
Pests/Disease: Mealybugs, spider mites, scale, fluoride toxicity, boron and calcium deficiencies.
Notes: Trim necrotic leaf tips to natural pattern (Figure A-43).
D. d. 'Lemon Lime' has yellow-green variegation on leaf margins (Figure A-44).

FIGURE A-44: *Dracaena deremensis* 'Lemon Lime', Lemon Lime Dracaena

Dracaena fragrans 'massangeana'

Pronunciation: dra-KEE-na FRAH-grans
Common name: Variegated Corn Plant
Family name: *Dracaenaceae*
Description: Tan-colored canes with terminal flush of long, recurved leaves having longitudinal swath of yellow-green variegation.

Maintenance: Bright, indirect light, tolerant of lower light levels; water when soil is dry to the touch; wipe dusty leaves with a soft cloth to reveal natural gloss.

Propagation: Stem tip cuttings, stem segment cuttings, air layering.

Pests/Disease: Mealybugs, spider mites, scale, fluoride toxicity, boron and calcium deficiencies.

Notes: *Fragrans* refers to this plant's fragrant, arching, terminal panicle of flowers, which can appear indoors under optimal, bright light conditions (Figure A-45).

FIGURE A-45: A cultivated variety of *Dracaena fragrans*, Variegated Corn Plant

Dracaena marginata

Pronunciation: dra-KEE-na mar-gi-NAH-ta

Common name: Red-Margined Dracaena, Dragon Palm

Family name: *Dracaenaceae*

Description: Linear, lance-shaped leaves atop gray stems, leaf margins variegated maroon red.

Maintenance: Bright, indirect light, tolerant of lower light levels; water when soil is dry to the touch; trim necrotic leaf tips.

Propagation: Stem tip cuttings, stem segment cuttings, air layering.

Pests/Disease: Mealybugs, spider mites, scale.

Notes: At lower light intensities, new growth is narrow, wispy, and curved while in higher levels it is broad and straight. Cutting of terminal growth may encourage branching in production.

Dracaena marginata 'tricolor'

Pronunciation: dra-KEE-na mar-gi-NAH-ta TRI-color

Common name: Rainbow Tree, Tricolor Dracaena

Family name: *Dracaenaceae*

Description: Linear, lance-shaped leaves atop gray stems, the leaves variegated green, cream, and pink.

Maintenance: Bright, indirect light, tolerant of lower light levels; water when soil is dry to the touch; trim necrotic leaf tips.

Propagation: Stem tip cuttings, stem segment cuttings, air layering.

Pests/Disease: Mealybugs, spider mites, scale.

Notes: Cutting of terminal growth may encourage branching in production. Higher light intensity increases variegation (Figure A-46).

FIGURE A-46: *Dracaena marginata 'Tricolor'*, Rainbow Tree, Tricolor Dracaena

FIGURE A-47: *Dracaena reflexa*, Dracaena reflexa, Pleomele

Dracaena reflexa

Pronunciation: dra-KEE-na re-FLEKS-a

Common name: Dracaena reflexa, Pleomele

Family name: *Dracaenaceae*

Description: Snaking canes with short, clasping, dark green foliage.

Maintenance: Bright, indirect light, tolerant of lower light levels; water when soil is dry to the touch.

Propagation: Stem tip cuttings, stem segment cuttings, air layering.

Pests: Mealybugs, spider mites, scale.

Notes: Synonymous with *Pleomele reflexa* (Figure A-47).

Dracaena reflexa 'Song of India' foliage is dark green with a longitudinal yellow stripe (Figure A-48).

FIGURE A-48: *Dracaena reflexa 'Song of India'*, Song of India Dracaena

Dracaena sanderiana

Pronunciation: dra-KEE-na sahn-da-ree-AH-na

Common name: Ribbon Plant, Lucky Bamboo

Family name: *Dracaenaceae*

Description: Cane-like stems with slightly rippled, lance-shaped, green leaves with longitudinal white stripes.

Maintenance: Bright, indirect light, tolerant of lower light levels; water when soil is dry to the touch. Cut back when leggy or propagate new plants via cuttings.

Propagation: Stem tip cuttings, stem segment cuttings, sow seeds.

Pests: Mealybugs, spider mites, scale.

Notes: Native to Cameroon. Canes of *D. sanderiana* used for Lucky Bamboo novelties (Figure A-49, Figure A-50).

FIGURE A-49: *Dracaena sanderiana*, Ribbon Plant

Dracaena surculosa

Pronunciation: dra-KEE-na sur-kew-LOW-sa

Common name: Gold Dust Dracaena

Family name: *Dracaenaceae*

Description: Small shrub at maturity, oval leaves having dark green ground, splotched with creamy yellow.

Maintenance: Bright, indirect light, tolerant of lower light levels; water when soil is dry to the touch. Cut back when leggy or propagate new plants via cuttings.

Propagation: Stem tip cuttings, sow seeds.

Pests: Mealybugs, spider mites, scale.

Notes: Often used as an accent plant in dish gardens (Figure A-51).

Echeveria sp.

Pronunciation: ih-chay-VAIR-ee-a

Common name: Echeveria

Family name: *Crassulaceae*

Description: Approximately 150 different species of succulents and subshrubs found in Texas, Mexico, and South America, foliage usually in rosettes in green, blue-green, and tinged with red or orange; flowers borne on long stalks from leaf axils.

Maintenance: Bright light, plant in sandy cactus mix, allow soil to dry between waterings.

Propagation: Pot up offsets, root stem, and leaf cuttings, or sow seeds.

Pests/Disease: Mealybugs, soft rot, leaf and stem rots.

Notes: Named after botanical artist Athanasio Echeverria Godoy (Figure A-52).

Epipremnum aureum

Pronunciation: e-pi-PREM-num OW-ree-um

Common name: Pothos, Devil's Ivy

Family name: *Araceae*

Description: Root-clinging climbers, alternate leaves, leaf morphology may differ on same plant as they mature.

Maintenance: Tolerant of low to high light levels, but variegation is most prominent with higher light. Stems may be supported or trained on numerous armatures/supports; pinch to remove leggy stems.

Propagation: Stem cuttings.

Pests/Disease: Spider mites, scale, leaf spot, root rot.

Notes: Synonymous with *Scindapsus aureus. Aureum* refers to the golden variegation. Sap may irritate skin (Figure A-53, Figure A-54).

Epipremnum pictum 'argyraeum'

Pronunciation: E. PIC-tum ar-jy-REE-um

Description: Dark green, heart-shaped leaves with variable silver spots (Figure A-55).

FIGURE A-50: *Dracaena sanderiana*, Lucky Bamboo

FIGURE A-51: *Dracaena surculosa*, Gold Dust Dracaena

FIGURE A-52: *Echeveria sp.*, Echeveria

FIGURE A-53: *Epipremnum aureum*, Pothos, Devil's Ivy

Eucharis amazonica

Pronunciation: EW-ca-ris a-ma-ZON-i-ca
Common name: Eucharis Lily
Family name: *Amaryllidaceae*
Description: Perennial bulb plants with broad, glossy, dark green, heart-shaped leaves; scapes bearing umbels of daffodil-like flowers with chartreuse green cups and white tepals.
Maintenance: Bright, filtered light, soil mix containing sharp sand and organic matter.
Propagation: Divide offsets.
Pests/Disease: Mealybugs.
Notes: *Eucharis* from the Greek meaning pleasing, charming.

FIGURE A-54: Upright Pothos trained on a wooden pole

Euphorbia candelabrum

Pronunciation: ew-FOR-bee-a can-del-AH-brum
Common name: Candelabra Euphorbia
Family name: *Euphorbiaceae*
Description: Oblong stem segments, mottled gray and green, freely branching succulent, paired thorns and short-lived, oval leaves.
Maintenance: Requires somewhat sandy soil, bright light, and little water.
Propagation: Root stem cuttings or sow seeds.
Pests/Disease: Spider mites, aphids, mealybugs, fungal and bacterial diseases.
Notes: Found in Somalia and South Africa (Figure A-56).

FIGURE A-55: *Epipremnum pictum* 'Argyraeum', Devil's Ivy

Euphorbia milii

Pronunciation: ew-FOR-bee-a MIL-ee-ee
Common name: Crown of Thorns
Family name: *Euphorbiaceae*
Description: Bushy, fleshy, thorny, gray-green stems, freely branching, bright green leaves, yellow cyathia enclosed by red involucres.
Maintenance: Requires somewhat sandy soil, bright light, and little water.
Propagation: Root stem cuttings or sow seeds.
Pests/Disease: Spider mites, aphids, mealybugs, fungal and bacterial diseases.

Notes: Native to Madagascar. Cultivars include pink, yellow, and orange involucres and variations of leaves and growth habits (Figure A-57).

Euphorbia pulcherrima

Pronunciation: ew-FOR-bee-a pul-KE-ri-ma
Common name: Poinsettia
Family name: *Euphorbiaceae*
Description: Partially deciduous shrub, leaves mid to deep green or variegated, terminal cymes bear yellow cyathia atop brightly colored involucral bracts in red, pink, white, peach, pale yellow, and bicolor.
Maintenance: Bright, indirect light; water when soil is dry to the touch. Avoid mechanical injury to branches. Sap may irritate skin.
Propagation: Stem cuttings.
Pests/Disease: Whiteflies, spider mites, aphids, mealybugs.
Notes: *Pulcherrima* in Latin translates as "very pretty"; many cultivars have been produced. Named for Joel Poinsett, statesman and First Minister (precursor to ambassadors) to Mexico, who introduced the plant to the United States around 1830 (Figure A-58).

Euphorbia tirucallii

Pronunciation: ew-FOR-bee-a ti-ru-CAL-ee-ee
Common name: Pencil Tree
Family name: *Euphorbiaceae*
Description: Fleshy, bright green, slender stems; linear, short-lived leaves; flowers consist of green cyathia and involucres.
Maintenance: Requires somewhat sandy soil, bright light, and little water.
Propagation: Stem cuttings.
Pests/Disease: Spider mites, aphids, mealybugs, fungal and bacterial diseases.
Notes: Sap may irritate skin (Figure A-59).

Exacum affine

Pronunciation: EKS-a-kum a-FEE-nee
Common name: Persian Violet
Family name: *Gentianaceae*

FIGURE A-56: *Euphorbia candelabrum*, Candelabra Euphorbia

FIGURE A-57: *Euphorbia milii*, Crown of Thorns

FIGURE A-58: *Euphorbia pulcherrima*, Poinsettia

FIGURE A-59: *Euphorbia tirucallii*, Pencil Tree

Description: Bushy perennial grown as an annual potted plant with small, bright green, shiny, oval leaves and violet, white, or rose star-shaped, fragrant flowers, .5 inches (1.25 centimeters) across.

Maintenance: Bright light, well-drained soil kept evenly moist but not wet.

Propagation: Seed.

Pests/Disease: Whiteflies, *Botrytis*, root rot.

Notes: Generally grown for springtime market.

X fatshedera lizei

Pronunciation: fats-HE-de-ra lee-ZAY-ee

Common name: Fatshedera, Tree Ivy

Family name: Araliaceae

Description: An intergeneric hybrid of the parent plants Fatsia and Hedera; leaves are palmately lobed, deep green; can climb if supported.

Maintenance: Tolerant of lower light conditions but does better with bright, indirect light. Use a soil-based potting mix, water when plant is dry to the touch.

Propagation: Vegetative cuttings.

Pests/Disease: Spider mites, mealybugs, scale, whiteflies, fungal and bacterial leaf spots.

Notes: *Fatsia japonica* 'Moseri' X *Hedera hibernica* raised by Lize Brothers of Nantes, France, 1910.

Fatsia japonica

FIGURE A-60: *Fatsia japonica*, Japanese Aralia, Japanese Fatsia

Pronunciation: FATS-e-a ja-PON-i-ka

Common name: Japanese Aralia, Japanese Fatsia

Family name: *Araliaceae*

Description: An evergreen perennial, leaves are palmately lobed, deep green, and leathery; can climb if supported.

Maintenance: Tolerant of lower light conditions but does better with bright, indirect light. Use a soil-based potting mix, water when plant is dry to the touch.

Propagation: Vegetative cuttings, air layering.

Pests/Disease: Spider mites, mealybugs, scale, whiteflies, fungal and bacterial leaf spots.

Notes: *Japonica*. Of Japan, also found in South Korea. Used as a potted plant and cut foliage (Figure A-60).

Ficus benjamina

Pronunciation: FEE-kus ben-ja-MEEN-a

Common name: Weeping Fig

Family name: *Moraceae*

Description: Shrub to tree, oval, elliptic, glossy leaves tapering to a point, indoor height to 20 feet (6 meters).

Maintenance: Bright, indirect light; water when soil is dry to the touch. Changes in light intensity cause leaf drop and must be anticipated when plants are moved or shifted.

Propagation: Stem cuttings, air layering.

Pests/Disease: Mealybugs, scale, spider mites, fungal and bacterial leaf spots, *Phomopsis*.

Notes: Native to South and Southeast Asia, North Australia, Southwest Pacific. Variegated, dwarf, and novelty foliage varieties exist. *Benjan* is the Indian name (Figure A-61).

FIGURE A-61: *Ficus benjamina*, Weeping Fig

Ficus binnendijkii 'amstel king'

Pronunciation: FEE-kus bin-nen-DIKE-ee-ee

Common name: Amstel King Ficus

Family name: *Moraceae*

Description: Oval-linear leaves, new growth tinged orange or red, brown bark bearing lenticels, indoor height to 20 feet (6 meters).

Maintenance: Bright, indirect light; water when soil is dry to the touch.

Propagation: Stem cuttings, air layering.

Pests/Disease: Mealybugs, scale, spider mites, fungal and bacterial leaf spots, *Phomopsis*.

Notes: Leaves of this species are slightly broader than *F. alii*.

Ficus deltoidea

Pronunciation: FEE-kus del-TOY-dee-a

Common name: Mistletoe Fig

Family name: *Moraceae*

Description: Shrub to tree, delta-shaped, glossy leaves, sometimes epiphytic, indoor height to 20 feet (6 meters).

Maintenance: Bright, indirect light; water when soil is dry to the touch.

FIGURE A-62: *Ficus deltoidea*, Mistletoe Fig

FIGURE A-63: *Ficus elastica*, Rubber Plant

FIGURE A-64: A variegated Rubber Plant cultivar

FIGURE A-65: *Ficus lyrata*, Fiddle-leaf Fig

Propagation: Stem cuttings, air layering.

Pests/Disease: Mealybugs, scale, spider mites, fungal and bacterial leaf spots.

Notes: Native to Southeast Asia, Borneo, and the Philippines (Figure A-62).

Ficus elastica

Pronunciation: FEE-kus e-LAS-ti-ka

Common name: Rubber Plant

Family name: *Moraceae*

Description: Shrub to tree, oval, elliptic, glossy leaves, 12 to 18 inches (30 to 45 centimeters) long, tapering to a point, indoor height to 20 feet (6 meters).

Maintenance: Bright, indirect light; water when soil is dry to the touch. Dusting the broad, glossy leaves is a part of maintenance.

Propagation: Stem cuttings, air layering.

Pests/Disease: Mealybugs, scale, spider mites, fungal and bacterial leaf spots, *Phomopsis*.

Notes: Variegated, dwarf, and novelty foliage varieties exist. Native to the Himalayas, Burma, and Java (Figure A-63, Figure A-64).

Ficus lyrata

Pronunciation: FEE-kus li-RAH-ta

Common name: Fiddle-leaf Fig

Family name: *Moraceae*

Description: Tree, leathery, glossy, lyre-shaped leaves, 10 to 18 inches (25 to 45 centimeters) long, indoor height to 20 feet (6 meters).

Maintenance: Bright, indirect light; water when soil is dry to the touch. Changes in light intensity cause leaf drop and must be anticipated when plants are moved.

Propagation: Stem cuttings, air layering.

Pests/Disease: Mealybugs, scale, spider mites, fungal and bacterial leaf spots.

Notes: Native to tropical West and Central Africa (Figure A-65).

Ficus maclellandii 'alii' ficus alii

Pronunciation: FEE-kus ma-KLEL-land-ee-ee ALL-ee-ee

Common name: Ficus Alii

Family name: *Moraceae*

Description: Linear, lance-shaped leaves, cinnamon brown bark bearing lenticels and numerous prop roots, indoor height to 20 feet (6 meters).

Maintenance: Bright, indirect light; water when soil is dry to the touch.

Propagation: Stem cuttings, air layering.

Pests/Disease: Mealybugs, scale, spider mites, fungal and bacterial leaf spots, *Phomopsis*.

Notes: Network of prop roots can be left in place for rugged effect or pruned for clean lines (Figure A-66).

FIGURE A-66: *Ficus maclellandii 'Alii'*, Ficus Alii

Ficus pumila

Pronunciation: FEE-kus PEW-mi-la

Common name: Creeping Fig

Family name: *Moraceae*

Description: Evergreen perennial, root-clinging climber, asymmetrical, heart-shaped leaves that are 1 inch (2.5 centimeters) long, but can be larger in older plants.

Maintenance: Bright, indirect light; do not allow soil to dry.

Propagation: Stem cuttings.

Pests/Disease: Mealybugs, scale, spider mites, fungal and bacterial leaf spots.

Notes: *Pumila* is Latin for dwarf. Plants are trained to moss-covered forms to create topiary. Native to China, Japan, and Vietnam (Figure A-67).

FIGURE A-67: *Ficus pumila*, Creeping Fig

Fittonia verschaffeltii

Pronunciation: fi-TON-ee-a vair-sha-FELT-ee-ee

Common name: Nerve Plant

Family name: *Acanthaceae*

Description: Evergreen perennial with mat-forming stems, opposite leaves with bright-colored veins.

Maintenance: Bright, indirect light; high humidity and moist, but not wet, soil.

Propagation: Stem cuttings.

Pests/Disease: Bacterial and fungal leaf spots, root rot.

Notes: Varying species have red or white veins. Good candidate for terrarium plant (Figure A-68).

FIGURE A-68: *Fittonia verschaffeltii*, Nerve Plant

FIGURE A-69: *Guzmania lingulata*, Guzmania

Guzmania lingulata

Pronunciation: guz-MAN-ee-a ling-gew-LAH-ta

Common name: Guzmania

Family name: *Bromeliaceae*

Description: Variable, epiphytic bromeliad, most often with deep green leaves, stems bear colorful bracts in red, yellow, orange, pink, light green, and more around tubular-shaped flowers.

Maintenance: Use a potting mix high in organic matter; keep soil evenly moist but not wet. Plants can be grown epiphytically or in pots with soil. Bright, indirect light keeps the bracts colorful for longer display periods.

Propagation: Pot offsets or sow seeds.

Pests/Disease: Mealybugs, fungal leaf spots.

Notes: *Lingulata* translates as tongue-like in reference to colorful bracts (Figure A-69).

FIGURE A-70: *Gynura aurantiaca*, Purple Velvet Plant

Gynura aurantiaca

Pronunciation: gin-EW-ra ow-ran-tee-AH-ka

Common name: Purple Velvet Plant

Family name: *Compositae*

Description: Erect, then running stems, green leaves ovate and coarsely toothed, covered in royal purple hairs, orange flowers have somewhat unpleasant odor and may be removed directing plant energy into vegetative production.

Maintenance: Bright filtered light; water before soil completely dries; stems may be pinched to encourage bushy growth.

Propagation: Stem cuttings.

Pests: Aphids, spider mites.

Notes: Good candidate for hanging baskets (Figure A-70).

Haworthia fasciata

Pronunciation: haw-WERTH-ee-a fas-SKI-a-ta

Common name: Haworthia

Family name: *Liliaceae*

Description: Clump-forming, stemless succulent, dark green leaves with white tubercles, stems to 16 inches (40 centimeters) bear small, funnel-shaped flowers.

Maintenance: Bright light, sandy cactus mix soil, low humidity, infrequent watering.

Propagation: Pot up offsets, division, seed, some species can be propagated by leaf cuttings.

Pests/Disease: Mealybugs, scale, soft rot.

Notes: Named for Adrian Hardy Haworth (1767–1833), British botanist. Native to South Africa (Figure A-71).

FIGURE A-71: *Haworthia fasciata*, Haworthia

Hedera helix

Pronunciation: HED-er-a HEE-liks

Common name: Ivy, English Ivy

Family name: *Araliaceae*

Description: Climbing or trailing perennial, 3- or 5-lobed triangular or broadly ovate leaves.

Maintenance: Bright, indirect light; keep soil evenly moist but not wet; provide humidity.

Propagation: Stem cuttings.

Pests/Disease: Spider mites, mealybugs, scale, aphids, bacterial spot, stem rot, and fungal leaf spots.

Notes: Native to Europe, a good candidate for terrarium plantings, hanging baskets, dish gardens, or specimen plants. Because the plant is frequently grown in 3- to 4-inch (7.5- to 10-centimeter) pots, care should be taken to provide moisture more frequently as roots could dry quickly, killing the plant. It is highly susceptible to spider mite infestation when displayed in warm, dry environments (Figure A-72).

FIGURE A-72: *Hedera helix*, Ivy, English Ivy

Hibiscus sp.

Pronunciation: hi-BIS-kus

Common name: Hibiscus

Family name: *Malvaceae*

Description: Bushy evergreen with oval to lance-shaped, deep green leaves; five-petaled flowers are produced from late spring until late fall.

Maintenance: Bright light; allow soil to dry slightly but not thoroughly between waterings. Under-watering can cause lower leaves to yellow and drop.

Propagation: Stem cuttings.

Pests/Disease: Whiteflies, aphids, mealybugs, scale, spider mites, stem and root rots.

Notes: Flowers are ephemeral, lasting for one day, and then abscising. Thus, they can cause hazardous debris on floors and walkways. Good container and bedding plant at exterior sites.

Hippeastrum sp.

FIGURE A-73: *Hippeastrum sp.,* Amaryllis

Pronunciation: hi-pee-OSS-trum

Common name: Amaryllis am-a-RIL-lis

Family name: *Amaryllidaceae*

Description: Bulb-forming perennials found in Central and South America, strap-shaped leaves develop in tandem with or just after flowering, trumpet-shaped flower clusters born on leafless stalks.

Maintenance: Plant bulbs with neck and shoulders above the soil in organic media, water thoroughly allowing soil to dry somewhat between waterings; provide bright, indirect light. After blooming, remove scape and grow as foliage plant, allowing it to naturally senesce, to provide energy for next year's bloom.

Propagation: Pot offsets, sow seed.

Pests/Disease: Bacterial and fungal rots, leaf spots, viruses.

Notes: Flowering can be slowed by keeping plants between 55 and 60 degrees F out of strong light (Figure A-73).

Homalomena sp.

Pronunciation: hom-a-low-MEE-na

Common name: Homalomena, King of Hearts

Family name: *Araceae*

Description: Heart-to arrowhead-shaped, dark green leaves on slender petioles, short stems, overall bushy habit, and slow grower.

Maintenance: Indirect light, tolerant of low light levels indoors. Broad foliage benefits from periodic cleaning with soft cloth.

Propagation: Division.

Pests/Disease: Mealybugs, scale.

Notes: Native to New Guinea.

Howea forsteriana

Pronunciation: HOW-ee-a for-ster-ee-AH-na

Common name: Kentia Palm

Family name: *Palmae*

Description: Slow-growing, single-stemmed palms with arching, somewhat pendulous fronds, dark green pinnae.

Maintenance: Tolerant of low light and neglect in comparison to most other palms.

Propagation: Seed.

Pests/Disease: Scale, spider mites, fungal leaf spots, root rot, manganese deficiency.

Notes: Due to lengthy production time, more costly, but more durable, than other palms (Figure A-74). Native to the porous, rocky soil of Lord Howe Island, Australia.

FIGURE A-74: *Howea forsteriana*, Kentia Palm

Hoya carnosa 'exotica'

Pronunciation: HOY-a kar-NO-sa

Common name: Exotic Hoya, Hoya

Family name: *Asclepiadaceae*

Description: Succulent climbing plant with ovate, fleshy leaves, variegated green, pink, and cream, waxy, star-shaped, light brown flowers borne in hemispherical to spherical umbels, chocolate-fragranced.

Maintenance: Bright light to filtered light; well-drained, organic potting mix.

Propagation: Root stem cuttings.

Pests/Disease: Mealybugs, scale.

Notes: Displays well in hanging baskets or stems can be trained to armatures. Native to India, southern China, and Burma (Figure A-75).

Hoya carnosa 'Hindu Rope' Hindu Rope has a similar growth habit, but with folded, curled, contorted leaves densely arranged on the stem.

FIGURE A-75: *Hoya carnosa 'Exotica'*, Exotic Hoya, Hoya

Hypoestes phyllostachya 'splash'

Pronunciation: hi-poe-ES-teez fil-low-STACK-ee-a

Common name: Splash Polka-Dot Plant

Family name: *Acanthaceae*

Description: Evergreen perennial, leaves are opposite, oval with reddish or pink spots, small spikes of flowers appear when light intensity is great.

Maintenance: Bright indirect light, medium to high humidity; pinch new growth to keep plants compact.

Propagation: Stem cuttings or seed.

Pests/Disease: Powdery mildew.

Notes: Height to 12 inches (30 centimeters), good for terrarium culture.

Iresine lindenii

Pronunciation: ee-res-EE-nay lin-DEN-ee-ee

Common name: Blood Leaf

Family name: *Amaranthaceae*

Description: Bushy perennial, ovate leaves, stems and leaves deep red, up to 24 inches (60 centimeters) indoors.

Maintenance: Bright light; water when soil is dry to the touch.

Propagation: Stem tip cuttings.

Pests/Disease: Aphids, spider mites, powdery mildew.

Notes: Native to Ecuador.

Iresine herbstii (I. HAIRBST-ee-ee), Beefsteak Plant, similar to *I. lindenii*, yellow, red, or orange leaves.

Juncus effusus 'spiralis'

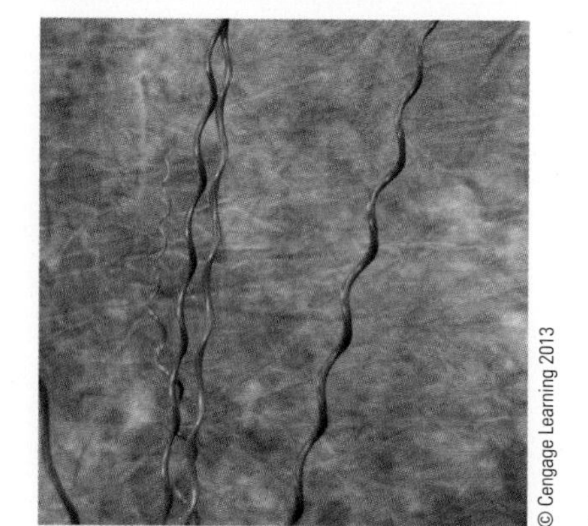

FIGURE A-76: *Juncus effusus* 'Spiralis', Corkscrew Rush

Pronunciation: YUNG-kus e-FEW-sus spee-RAH-lis

Common name: Corkscrew Rush

Family name: *Juncaceae*

Description: Grass-like perennial; dark green, leafless, spiraling stems.

Maintenance: Bright light, evenly moist soil.

Propagation: Division or seed.

Pests/Disease: Fungal leaf spot, stem rot.

Notes: Novel form of leaves interesting as potted plant and cut foliage (Figure A-76).

Justicia brandegeana

Pronunciation: JOOS-tis-ee-a brand-ee-gee-AH-na

Common name: Shrimp Plant

Family name: *Acanthaceae*

Description: Evergreen shrub, downy stems and flower bracts which resemble shrimp in form, white flowers.

Maintenance: Bright, indirect light; prune leggy growth to keep plant bushy.

Propagation: Stem cuttings or seed.

Pests/Disease: Spider mites, whiteflies, bacterial and fungal leaf spots.

Notes: Native to Mexico. Named for James Justice (1698–1763), Scottish botanist.

Kalanchoe blossfeldiana

Pronunciation: ka-LAN-koe-ee blos-feld-ee-AH-na

Common name: Kalanchoe

Family name: *Crassulaceae*

Description: Succulent oval to ovate leaves, scalloped margins, deep green, long stalks bear masses of four-petaled flowers, various cultivars with flowers in red, orange, yellow, white, pink, and lavender.

Maintenance: Bright light, soil-based potting mix with sand or grit for drainage. Higher light intensities indoors will prolong bloom duration.

Propagation: Stem cuttings, pot offsets, or sow seed.

Pests/Disease: Mealybugs, aphids, powdery mildew, leaf spots, crown rot, root rot, bacterial soft rot.

Notes: Named after German nurseryman Robert Blossfeld, Pottsdam, Germany, active 1920–1940. Often used for color rotations in interiorscapes and can be used in exterior plantings, re-flowering within 6 to 8 weeks in full sun (Figure A-77).

FIGURE A-77: *Kalanchoe blossfeldiana*, Kalanchoe

Kalanchoe daigremontiana

Pronunciation: ka-LAN-koe-ee day-gre-mont-ee AH-na

Common name: Devil's Backbone, Mother of Thousands

Family name: *Crassulaceae*

Description: Succulent lanceolate leaves, adventitious plantlets developing on toothed margins, light green leaves speckled rosy-purple.

Maintenance: Bright light, soil-based potting mix with sand or grit for drainage. Higher light intensities indoors will prolong bloom duration.

Propagation: Pot offsets, stem cuttings.

Pests/Disease: Mealybugs, aphids, powdery mildew, leaf spots, crown rot, root rot, bacterial soft rot.

Notes: Cultivate as a houseplant due to interesting plantlets and ease of propagation (Figure A-78).

FIGURE A-78: *Kalanchoe daigremontiana*, Devil's Backbone, Mother of Thousands

Maranta leuconeura

Pronunciation: ma-RAN-ta loo-co-NEWR-ra

Common name: Prayer Plant

Family name: *Marantaceae*

Description: Clumping perennial, oval, dark green leaves with red veins flaring from midrib to margins, leaf undersides light maroon; small, white, tubular flowers produced on small racemes.

Maintenance: Bright light to filtered light, shallow pots accommodate fine root system, thus keeping plant from drying out; rich, well-drained soil; high humidity.

FIGURE A-79: *Maranta leuconeura*, Prayer Plant

Propagation: Division, cuttings, seed.

Pests/Disease: Mealybugs, spider mites, leaf spots.

Notes: Leaves characteristically spread to accumulate light energy during the day, then fold upright at night. Well-displayed in hanging baskets (Figure A-79).

Maranta leuconeura var. Kerchoviana (M. l. var. ker-chov-ee-AH-na), Rabbit's Tracks, has bright green leaves with coffee brown marks and pale green undersides.

Monstera deliciosa

FIGURE A-80: *Monstera deliciosa*, Split Leaf Philodendron

Pronunciation: mon-STEH-ra day-li-KEE-oh-sa

Common name: Split Leaf Philodendron

Family name: *Araceae*

Description: Thick-stemmed climber produces pinnatifid and perforated mature leaves while juvenile leaves are entire; spathes produce edible fruit. Variegated cultivars exist.

Maintenance: Bright, indirect light with high humidity, soil high in organic matter; may require pruning in growth-type situations.

Propagation: Stem tip cuttings, stem segment cuttings with adventitious root presence, leaf cuttings, air layering.

Pests/Disease: Mealybugs, scale, spider mites, bacterial soft rot, and leaf spots.

Notes: Native to southern Mexico to Panama (Figure A-80).

Neoregelia sp.

FIGURE A-81: *Neoregelia sp.*, Neoregelia

Pronunciation: nee-oh-ray-GELL-ee-a

Common name: Neoregelia

Family name: *Bromeliaceae*

Description: Rhizomatous or stoloniferous, terrestrial or epiphytic perennial bromeliads. Leaf rosettes totally enclose short scape and bracts.

Maintenance: Bright, indirect light; pot in well-drained, organic bromeliad mix; keep water in leafy vase.

Propagation: Pot offsets keeping vase and rosette base above soil line to prevent rot.

Pests: Scale, mealybugs, bacterial soft rot, leaf spot.

Notes: Nestled scape is a key in discerning *Neoregelia* from other *Bromeliads* (Figure A-81).

Nephrolepis exaltata

Pronunciation: nef-ROE-lep-is eks-al-TAH-ta

Common name: Boston Fern

Family name: *Oleandraceae*

Description: The source of nearly all *Nephrolepis* cultivars, this fern has erect fronds when juvenile, arching fronds when mature to 6 feet (11 meters) long, fang-shaped pinnae following a central rachis, the overall plant is cushion-shaped, numerous running rhizomes.

Maintenance: Bright, indirect light; allow soil to be evenly moist but not wet, humid conditions; does well in organic, well-drained soil.

Propagation: Division, root runners, sow spores.

Pests/Disease: Spider mites, scale, mealybugs, leaf spots, root rots.

Notes: Seen in Florida, Mexico, West Indies, Central America, tropical South America, Polynesia, and Africa.

Nertera granadensis

Pronunciation: NER-te-ra gran-a-DEN-sis

Common name: Bead Plant

Family name: *Rubiaceae*

Description: Moss-like perennial plant with small, round, bright green leaves 3/8 inch ((9 millimeters) long, yellow flowers followed by bright orange, round fruit.

Maintenance: Bright, filtered light in organic potting mix; do not allow to dry out completely.

Propagation: Division or seed.

Pests/Disease: Spider mites, aphids.

Notes: *Nerteros*, from Greek, (lowly) in reference to dwarf habit.

Oncidium sp.

Pronunciation: on-KID-ee-um

Common name: Oncidium

Family name: *Orchidaceae*

Description: Most varieties used for interior plants have pseudobulbs bearing leaves and producing tall racemes from the base of the plant, bright yellow flowers bearing prominent lips.

Maintenance: Bright, filtered light; pot in orchid mix or interior plant soil mixed with bark, allow to slightly dry between waterings.

Propagation: Division, tissue culture.

Pests/Disease: Aphids, mealybugs, spider mites, whiteflies, viruses, *Botrytis*.

Notes: Native to numerous habitats in Mexico, Central America, South America, and West Indies.

Pachypodium lamerei

Pronunciation: pa-kee-PODE-ee-um la-MARE-ee
Common name: Pachypodium
Family name: *Apocynaceae*
Description: Succulents, often with thick caudices and thorny stems; terminal, lance-shaped leaves; broad, white flowers with yellow throats in summer.
Maintenance: Bright light; water sparingly and pot in a well-drained, cactus potting mix.
Propagation: Stem cuttings or seed.
Pests/Disease: Aphids, leaf spots, stem rot.
Notes: Native to Namibia, Madagascar, and South Africa.

Paphiopedilum sp.

Pronunciation: pa-fee-oh-PEH-di-lum
Common name: Lady Slipper Orchid, Slipper Orchid.
Family name: *Orchidaceae*
Description: Monopodial, short stems with strap-like leaves, ovate to lance-shaped, lacking in pseudobulbs, shoots typically bearing one flower having an upright sepal, two spreading petals, and two sepals under a prominent, pouch-shaped lip.
Maintenance: Use a terrestrial orchid potting mix, bright, filtered light; keeping the plant somewhat root-bound encourages blooming; do not allow soil to dry out between waterings.
Propagation: Separate newly formed plants from mature plants, tissue culture.
Pests/Disease: Spider mites, whiteflies, aphids, mealybugs, *Botrytis*, anthracnose, root rot, bacterial soft rot.
Notes: Found in India, China, Southeast Asia, New Guinea; from the Greek words *Paphos*, a worship site for Venus, and *pedilon*, footwear.

Pedilanthus tithymaloides

Pronunciation: pe-di-LANTH-us ti-thee-ma-LOI-dees
Common name: Devil's Backbone
Family name: *Euphorbiaceae*

Description: Succulent shrub to small tree, leaves are quickly deciduous, light green with white and pink variegation and borne on newer growth, stems are slightly zigzag in habit and exude a milky sap that can irritate skin.

Maintenance: Bright light; allow plant to dry between waterings.

Propagation: Stem tip cuttings, seed.

Pests/Disease: Powdery mildew, leaf spots.

Notes: Native to rocky terrain in Mexico, Central and South America (Figure A-82).

FIGURE A-82: *Pedilanthus tithymaloides*, Devil's Backbone

Peperomia caperata 'emerald ripple'

Pronunciation: pe-pe-ROM-ee-a ka-pe-RAH-ta

Common name: Emerald Ripple Peperomia

Family name: *Piperaceae*

Description: Perennial, mound forming with dark green, heart-shaped, rugose (rippled) leaves, spadix blooms.

Maintenance: Bright, indirect light, water when soil is dry to the touch, use soilless or soil-based potting mix.

Propagation: Separate offsets, seed.

Pests/Disease: Bacterial soft rot.

Notes: Textural houseplant, native to Brazil (Figure A-83).

FIGURE A-83: A cultivar of *Peperomia caperata*

Peperomia clusiifolia 'variegata'

Pronunciation: pe-pe-ROM-ee-a cloo-see-i-FO-li-a

Common name: Red-Edge Peperomia

Family name: *Piperaceae*

Description: Upright perennial with succulent stems, leaves obovate, green, tinged with pinkish red and cream variegation, spadix blooms.

Maintenance: Bright, indirect light, water when soil is dry to the touch, use soilless or soil-based potting mix.

Propagation: Stem cuttings.

Pests/Disease: Bacterial soft rot.

Notes: Native to Brazil (Figure A-84).

FIGURE A-84: *Peperomia clusiifolia 'Variegata'*, Red-Edge Peperomia

Peperomia obtusifolia peperomia

Pronunciation: pe-pe-ROM-ee-a

Common name: Peperomia, Baby Rubber Plant

Family name: *Piperaceae*

© Cengage Learning 2013

FIGURE A-85: *Peperomia obtusifolia*, Peperomia, Baby Rubber Plant

© Cengage Learning 2013

FIGURE A-86: *Peperomia obtusifolia* 'Variegata' Peperomia, Variegated Baby Rubber Plant

© Cengage Learning 2013

FIGURE A-87: *Phalaenopsis sp.*, Phalaenopsis, Butterfly Orchid, Moth Orchid

Description: Upright perennial with succulent stems, leaves oval to obovate, deep green, spadix blooms.

Maintenance: Bright, indirect light; water when soil is dry to the touch, use soilless or soil-based potting mix.

Propagation: Stem cuttings.

Pests/Disease: Bacterial soft rot.

Notes: Native to Brazil (Figure A-85, Figure A-86).

Peperomia scandens

Pronunciation: pe-pe-ROM-ee-a SKAN-dens

Common name: False Philodendron

Family name: *Piperaceae*

Description: Trailing stems, heart-shaped green leaves, spadix flowers.

Maintenance: Bright, indirect light; water when soil is dry to the touch, use soilless or soil-based potting mix. Pinch back leggy growth.

Propagation: Stem cuttings.

Pests/Disease: Bacterial soft rot.

Notes: Mexico to South America.

Phalaenopsis sp.

Pronunciation: fa-lie-NOP-sis

Common name: Phalaenopsis, Butterfly Orchid, Moth Orchid

Family name: *Orchidaceae*

Description: Monopodial, epiphytic, without pseudobulbs, with obovate to oval, fleshy green leaves, simple or branching racemes with showy flowers in pink, white, and other colors, potentially holding flowers for months.

Maintenance: Bright, indirect light, epiphytic orchid mix, drench media and allow for aeration and partial drying, able to re-flower on same branch.

Propagation: Cuttings, offshoots, tissue culture.

Pests/Disease: Bacterial soft rot, viruses.

Notes: From Greek *phalaina* (a moth) and *–opsis* (resembling) (Figure A-87).

Philodendron bipinnatifidum

Pronunciation: fi-lo-DEN-dron bi-pi-ni-TIF-i-dum

Common name: Tree Philodendron

Family name: *Araceae*

Description: Large single stem, recumbent with age, leaves borne on long petioles, deep green, to 3 feet (1 meter), deeply pinnatisect with wavy margins, spath and spadix blooms.

Maintenance: Medium to bright filtered light; allow soil mix to dry somewhat between waterings.

Propagation: Stem tip, air layer, sow seed.

Pests/Disease: Mealybugs, scale, spider mites, fungal and bacterial leaf spots, root rot.

Notes: Leaves are grown as a cut foliage crop (Figure A-88).

FIGURE A-88: *Philodendron bipinnatifidum*, Tree Philodendron

Philodendron domesticum

Pronunciation: fi-lo-DEN-dron do-MES-ti-kum

Common name: Spade-leaf Philodendron

Family name: *Araceae*

Description: Climbing, with shovel-shaped leaves, varieties are green or deep red.

Maintenance: Medium to bright filtered light; train on decorative pole if upright plant form is desired.

Propagation: Stem tip and stem segment including node.

Pests/Disease: Mealybugs, scale, spider mites, fungal and bacterial leaf spots, root rot.

Notes: Broad, shiny leaves provide contrast to smaller-leaved indoor plants (Figure A-89).

FIGURE A-89: *Philodendron domesticum*, Spade-leaf Philodendron

Philodendron micans

Pronunciation: fi-lo-DEN-dron MEE-cans

Common name: Micans Philodendron

Family name: *Araceae*

Description: Freely running perennial, heart-shaped, bronze-green leaves with reddish undersides, more velvety textured than *P. scandens*.

Maintenance: Fast-growing, keep leggy growth cut to encourage fullness, tolerant of low to medium light levels, water when soil becomes dry to touch.

Propagation: Stem tip and stem segment cuttings with nodes.

Pests/Disease: Mealybugs, scale, spider mites, fungal and bacterial leaf spots, root rot.

Notes: Easy to grow houseplant (Figure A-90).

FIGURE A-90: *Philodendron micans*, Micans Philodendron

FIGURE A-91: *Philodendron scandens*, Heartleaf Philodendron

Philodendron scandens

Pronunciation: fi-lo-DEN-dron SKAN-dens
Common name: Heartleaf Philodendron
Family name: *Araceae*
Description: Freely running perennial, heart-shaped, deep green, glossy leaves.
Maintenance: Fast-growing, keep leggy growth cut to encourage fullness, tolerant of low to medium light levels, water when soil becomes dry to touch.
Propagation: Stem tip and stem segment cuttings with nodes.
Pests/Disease: Mealybugs, scale, spider mites, fungal and bacterial leaf spots, root rot.
Notes: Easy to grow houseplant native to Mexico, West Indies, and Southeast Asia (Figure A-91).

FIGURE A-92: *Phoenix roebelenii*, Pygmy Date Palm

Phoenix roebelenii

Pronunciation: FEE-niks roe-bel-EN-ee-ee
Common name: Pygmy Date Palm
Family name: *Arecaceae*
Description: Single or clustering, small palm, to 6 feet (2 meter) tall, leaves 3 to 4 feet (1 to 1.2 meters) long, with glossy leaflets, with stiff hairs on lower side of rachis.
Maintenance: Bright light, soil-based potting mix, water thoroughly.
Propagation: Seed.
Pests/Disease: Spider mites, mealybugs, scale, thrips, leaf spots, and various rots.
Notes: Leaves are grown as a cut foliage crop. Native to Laos (Figure A-92).

Pilea cadierei

Pronunciation: PEE-lee-a ka-dee-EH-ree-ee
Common name: Aluminum Plant
Family name: *Urticaceae*
Description: Upright perennial, obovate, lance-shaped leaves are dark green with raised silver patterns; height to 12 inches (30 centimeters).
Maintenance: Bright, indirect light; water when soil is dry to the touch, soilless potting mix, pinch back to encourage bushiness.
Propagation: Stem tip cuttings or seed.

Pests/Disease: Spider mites, mealybugs, stem rot, bacterial leaf spot.

Notes: Named for R. P. Cadiere who collected the plant in Vietnam, 1938 (Figure A-93).

© Cengage Learning 2013

FIGURE A-93: *Pilea cadierei*, Aluminum Plant

Pittosporum tobira

Pronunciation: pi-TOSS-po-rum to-BYE-ra

Common name: Japanese Pittosporum

Family name: *Pittosporaceae*

Description: Small tree to shrub, light brown wood stems bearing terminal rosettes of deep green, leathery, oval leaves.

Maintenance: Use soil-based media, full light, allow to dry somewhat between waterings.

Propagation: Cuttings, air layering, sow seeds.

Pests/Disease: Mealybugs, spider mites, aphids, scale.

Notes: *Pittosporum tobira 'Variegata'*, Variegated Japanese Pittosporum has lighter green leaves and cream variegation. Both are grown and harvested as a cut foliage (Figure A-94).

© Cengage Learning 2013

FIGURE A-94: *Pittosporum tobira 'Variegata'*, Variegated Japanese Pittosporum

Platycerium bifurcatum

Pronunciation: pla-tee-KEH-ree-um bi-fur-KAH-tum

Common name: Staghorn Fern

Family name: *Polypodiaceae*

Description: Epiphytic fern with rounded sterile fronds that are pale green when juvenile, turning tan and spongy when mature; fertile fronds are green, spreading or hanging, in two to three lobes of strap-like segments covered in fine hairs.

Maintenance: Attach to board or branch and hang vertically, provide bright, indirect light, water freely, allowing water to penetrate sterile frond mass. Larger plants do well in space where water splash is not a problem.

Propagation: Separate plantlets or sow spores.

Pests/Disease: Scale, mealybugs.

Notes: *Platys* (broad), *keras* (horn) *bifurcatum* (forked in two) in reference to the leaf morphology (Figure A-95).

© Cengage Learning 2013

FIGURE A-95: *Platycerium bifurcatum*, Staghorn Fern

Plectranthus australis

Pronunciation: plek-TRANTH-us ow-STRAH-lis

Common name: Swedish Ivy

Family name: *Labiatae*

Description: Upright when juvenile, then hanging, square stems, with round, scalloped-margined, bright green, shiny leaves; terminal racemes of small, white flowers.

Maintenance: Soil-based potting mix; bright, indirect light, water when soil is dry to touch.

Propagation: Stem cuttings, division.

Pests/Disease: Mealybugs, spider mites, leaf spots, root rot.

Notes: From Australia; well suited as a suspended, hanging basket (Figure A-96).

FIGURE A-96: *Plectranthus australis*, Swedish Ivy

Plectranthus forsteri 'marginatus'

Pronunciation: plek-TRANTH-us FOR-ste-ree mar-gin-AH-tus

Common name: Variegated Swedish Ivy

Family name: *Labiatae*

Description: Upright when juvenile, then hanging, square stems, with round, scalloped-margined, velvety, light green leaves with white variegation on the margins, terminal racemes of small, white flowers; fragrant when crushed.

Maintenance: Soil-based potting mix; bright, indirect light, water when soil is dry to touch.

Propagation: Stem cuttings, division.

Pests/Disease: Mealybugs, spider mites, leaf spots, root rot.

Notes: Native to Australia, Fiji, and New Caledonia; well suited as a hanging basket.

Podocarpus (afrocarpus) gracilior

Pronunciation: pod-o-KAR-pus gra-KIL-ee-or

Common name: Weeping Podocarpus

Family name: *Podocarpaceae*

Description: Shrub to small tree, lance-shaped, leathery, dark green leaves, weeping, pendulous branches.

Maintenance: Bright light; water when soil is dry to the touch; use soil-based media.

Propagation: Cuttings, seed.

Pests/Disease: Mealybugs, scale.

Notes: Evergreen conifer.

Podocarpus (afrocarpus) macrophyllus

Pronunciation: pod-o-KAR-pus mak-roe-FIL-us

Common name: Podocarpus, Buddhist Pine

Family name: *Podocarpaceae*

Description: Shrub to small tree, lance-shaped, leathery, dark green leaves.

Maintenance: Bright light; water when soil is dry to the touch; use soil-based media.

Propagation: Cuttings, seed.

Pests/Disease: Mealybugs, scale.

Notes: Evergreen conifer.

FIGURE A-97: *Polyscias balfouriana 'Marginata',* Variegated Balfour Aralia

Polyscias balfouriana

Pronunciation: po-LIS-kee-as BAL-for-i-ah-na

Common name: Variegated Balfour Aralia

Family name: *Araliaceae*

Description: Upright evergreen shrub, bearing rounded leaves with scalloped margins.

Maintenance: Bright filtered light; grow in soil-based potting mix; may need pruning in growth-type situations.

Propagation: Stem tip and stem-segment cuttings, seed.

Pests: Mealybugs, scale, spider mites.

Notes: *P. b. 'Marginata'* has cream-variegated leaf margins (Figure A-97).

FIGURE A-98: *Polyscias filicifolia,* Ming Aralia

Polyscias filicifolia

Pronunciation: po-LIS-kee-as fi-li-ki-FO-li-a

Common name: Ming Aralia

Family name: *Araliaceae*

Description: Upright evergreen shrub, bearing bright green pinnate to 3-pinnate leaves composed of numerous leaflets, giving the effect of parsley foliage.

Maintenance: Bright filtered light; grow in soil-based potting mix; may need pruning in growth-type situations.

Propagation: Stem tip and stem-segment cuttings, seed.

Pests: Mealybugs, scale, spider mites (Figure A-98).

Radermachera sinica

Pronunciation: rod-er-MOCK-er-a SIN-i-ca

Common name: China Doll

Family name: *Bignoniaceae*

Description: Evergreen perennial, columnar indoors, with bipinnate, glossy green leaves made up of numerous leaflets.

FIGURE A-99: *Radermachera sinica*, China Doll

Maintenance: Well-drained soil mix; do not allow plant to dry between waterings.

Propagation: Stem tip cuttings, seed.

Pests/Disease: Aphids, mealybugs, scale.

Notes: Dry soil promotes leaf drop (Figure A-99).

Rhapis excelsa

Pronunciation: RA-pis eks-KEL-sa

Common name: Lady Palm

Family name: *Arecaceae*

Description: Cluster-stemmed palms with deep green, palmate leaves divided nearly to the leaf base into 4 to 10 lance-shaped, furrowed lobes. As an interior plant, available from 3 to 15 feet (1 to 4.5 meters) tall.

Maintenance: Bright, indirect light, soilless mix; water thoroughly, and then allow to dry somewhat between waterings.

Propagation: Division, seed.

Pests/Disease: Spider mites, scale, leaf spots.

Notes: Shady tropical forest habitats in Southeast Asia and southern China (Figure A-100).

FIGURE A-100: *Rhapis excelsa*, Lady Palm

Rosmarinus officinalis

Pronunciation: rose-ma-REEN-us o-fi-ki-NAL-is

Common name: Rosemary

Family name: *Labiatae*

Description: Evergreen shrubs with fragrant, needle-like, opposite leaves, used as culinary herb.

Maintenance: Bright light, well-drained soil containing sand; allow soil to dry somewhat between waterings.

Propagation: Vegetative cuttings, seed.

Pests/Disease: Bacterial leaf spots, root rot.

Notes: Native to the Mediterranean. Grown as an interior accent plant, cut foliage used in floral designs (Figure A-101).

FIGURE A-101: *Rosmarinus offincinalis*, Rosemary

Saintpaulia ionantha

Pronunciation: saynt-PAWL-ee-a ee-on-ANTH-a

Common name: African Violet

Family name: *Gesneriaceae*

Description: Fast-growing evergreen perennials, short stems, rosettes of oval, succulent, hairy leaves on long petioles; showy, attractive flowers, often five-petaled. Many variations of the plant are available.

Maintenance: Bright, indirect light encourages blooming; use a well-drained, soilless potting mix and avoid water splash on leaves; remove spent blossoms.

Propagation: Leaf cuttings, plantlets.

Pests/Disease: Aphids, mealybugs, thrips, cyclamen mites, spider mites, *Botrytis*, crown rot, powdery mildew.

Notes: Rewarding houseplant, *Saintpaulia* perform well under artificial lighting.

Sansevieria trifasciata

FIGURE A-102: *Sansevieria trifasciata 'Laurentii',* Snake Plant

Pronunciation: san-sev-ee-AY-ree-a tri-fas-ee-AH-ta

Common name: Snake Plant, Mother-in-Law's Tongue

Family name: *Agavaceae*

Description: Upright perennial, rhizomatous, with pointed, lance-shaped, fleshy leaves, with horizontal bands of light and dark green.

Maintenance: Bright filtered light, well-drained potting mix; transplant only when plant is pot-bound; allow to dry somewhat between waterings.

Propagation: Division, leaf segment cuttings. Variegated cultivars' offspring will lack variegation if rooted from leaves.

Pests/Disease: Mealybugs and spider mites.

Notes: Native to Africa, Madagascar, India, and Indonesia. *Trifasciata* means the flowers are grouped in threes. The Prince of Sanseviero, Italy, Raimond de Sangro, was an eighteenth-century horticulture benefactor.

Sansevieria trifasciata 'Laurentii' (S. t. lo-RENT-ee-ee), Snake Plant similar in growth and appearance to S.t., but has broad creamy-yellow, longitudinal stripes on margins (Figure A-102).

Sansevieria trifasciata 'Hahnii' (S. t. HAHN-ee-ee), Bird's Nest Sansevieria is a dwarf version with rosettes of leaves variegated green and dark green (Figure A-103).

FIGURE A-103: *Sansevieria trifasciata 'Hahnii'*, Bird's Nest Sansevieria

Saxifraga stolonifera

Pronunciation: sacks-IF-ra-ga sto-low-NI-fe-ra

Common name: Strawberry Begonia, Mother of Thousands

FIGURE A-104: *Saxifraga stolonifera*, Strawberry Begonia, Mother of Thousands

Family name: *Saxifragaceae*

Description: Rosette to tuft-forming perennial, plantlets produced from stolons; round, deep green leaves above and maroon below, with fine hairs on both sides.

Maintenance: Bright, indirect light; water when soil is dry to the touch.

Propagation: Root plantlets as individual plants.

Pests/Disease: Spider mites, aphids.

Notes: Houseplant or underplanting in interiorscapes. May be displayed as hanging plant to better observe stolons and plantlets. Native to East Asia (Figure A-104).

Schefflera actinophylla (syn. with brassaia actinophylla)

Pronunciation: shef-LE-ra ak-tin-oh-FIL-la

Common name: Schefflera

Family name: *Araliaceae*

Description: Shrub to small tree with leaves 6 to 14 inches (12 to 35 centimeters) across consisting of 7 to 16 obovate leaflets borne in terminal rosettes. Juvenile plants have fewer leaflets than mature plants.

Maintenance: Bright, indirect light.

Propagation: Vegetative cuttings, air layering, sow seed.

Pests: Scale, thrips, mealybugs.

Notes: Native to Southeast Asia, the Pacific Islands, Central and South America.

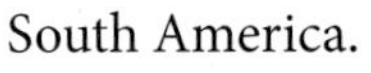

FIGURE A-105: *Schefflera arboricola*, Dwarf Schefflera

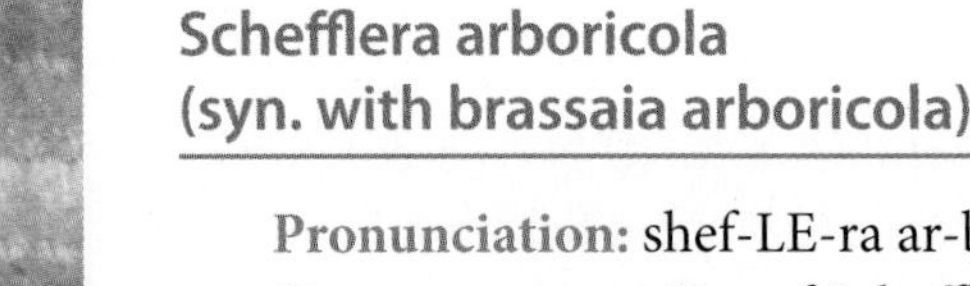

Schefflera arboricola (syn. with brassaia arboricola)

Pronunciation: shef-LE-ra ar-bo-RI-ko-la

Common name: Dwarf Schefflera

Family name: *Araliaceae*

Description: Shrub to small tree with leaves 4 to 8 inches (10 to 20 centimeters) long with 7 to 11 obovate leaflets.

Maintenance: Bright, indirect light.

Propagation: Vegetative cuttings, air layering, sow seed.

Pests: Scale, thrips, mealybugs.

Notes: Native to Taiwan, *Arboricola* translates as growing on trees (Figure A-105).

Schefflera elegantissima (syn. with dizygotheca elegantissima)

Pronunciation: shef-LE-ra *ay-le-gan-TIS-i-ma*
Common name: False Aralia
Family name: *Araliaceae*
Description: Shrub to small tree with purple-green, dissected, palmately compound borne in terminal rosettes. Habitat is upright and columnar indoors.
Maintenance: Bright, indirect light; water when soil is dry to the touch, kept evenly moist but not wet.
Propagation: Vegetative cuttings, air layering, sow seed.
Pests: Spider mites, scale, thrips, mealybugs.
Notes: Native to New Caledonia (Figure A-106).

FIGURE A-106: *Schefflera elegantissima,* False Aralia

Schlumbergera truncata (syn. zygocactus truncata)

Pronunciation: shlum-BER-ge-ra trun-KA-ta
Common name: Thanksgiving Cactus
Family name: *Cactaceae*
Description: Oblong, green stem segments with toothed margins, showy flowers in pink, white, red, or orange in late fall.
Maintenance: Bright light, cactus potting mix, allow plant to dry somewhat between waterings.
Propagation: Stem segment cuttings.
Pests/Disease: Mealybugs, leaf spots.
Notes: Native to Brazil; many cultivars.

Selaginella kraussiana

Pronunciation: se-lah-gi-NEL-la krows-ee-AH-na
Common name: Trailing Spikemoss, Selaginella
Family name: *Sellaginellaceae*
Description: Evergreen perennial forming a mat of bright green, pinnatisect foliage; about 1 inch (2.5 centimeters) in height, trailing habit carpets the soil surface.
Maintenance: Bright, indirect light, organic potting mix, do not allow media to dry completely, high humidity.
Propagation: Division.
Pests/Disease: Leaf spots, root rot.
Notes: A good terrarium plant and often the only way to enjoy display longevity (Figure A-107).

FIGURE A-107: *Selaginella kraussiana,* Trailing Spikemoss, Selaginella

Sinningia speciosa

Pronunciation: si-NING-ee-a spe-KEE-oh-sa
Common name: Florist Gloxinia
Family name: *Gesneriaceae*
Description: A rosette of ovate, velvety green leaves and scalloped margins arise from a tuber; bell-shaped, velvety flowers are borne at the crown of the plant in red, lavender, violet, white, or bicolor.
Maintenance: Bright light as from an eastern exposure, keep organic media moist but not overwatered.
Propagation: Pot offsets, leaf cuttings.
Pests/Disease: Cyclamen mites, aphids, whiteflies, thrips, leaf miners, crown rot, viruses.
Notes: Native to Brazil, plants perform well under artificial light.

Soleirolia soleirolii

Pronunciation: so-lay-ROL-ee-a so-lay-ROL-ee-ee
Common name: Baby's Tears, Mind Your Own Business
Family name: *Urticaceae*
Description: Mossy in appearance, light green, dense, carpet-forming dwarf perennial.
Maintenance: Bright, indirect light; keep soil evenly moist but not wet.
Propagation: Division
Pests/Disease: Uncommon.
Notes: Good candidate for terrarium planting. Because the plant is frequently grown in 3- to 4-inch (7.5- to 10-centimeter) pots, care should be taken to provide moisture more frequently as roots could dry out, quickly killing the plant.

Spathiphyllum sp.

Pronunciation: spa-thi-FIL-um
Common name: Peace Lily
Family name: Araceae
Description: Lance-shaped leaves, flowers bearing a white spathe and green or white spadices, the spathe turning green with age.
Maintenance: Most tolerant flowering plant for low light interiors; grow in soilless potting mix, keep evenly moist but not wet; plant shows water stress with downturned leaves.
Propagation: Division, sow seed.

Pests/Disease: Bacterial soft rot, leaf spots, and root rot.

Notes: Various cultivars are dwarf, variegated, larger-leaved, or more floriferous (Figure A-108).

FIGURE A-108: *Spathiphyllum sp.*, Peace Lily

Strelitzia reginae

Pronunciation: stre-LITS-ee-a ray-GEEN-ee

Common name: Bird of Paradise

Family name: Strelitziaceae

Description: Clump-forming perennials with lance-shaped leaves and long petioles, gray-green, may reach 30 feet (10 meters) in nature; inflorescences consist of horizontal, green spathes from which emerge orange-petaled flowers.

Maintenance: Bright light, allow soil mix to dry somewhat between waterings. Pot-bound plants have a better chance for flowering indoors.

Propagation: Divide rooted suckers, seed propagation.

Pests/Disease: Scale, fungal and bacterial spots.

Notes: *S. alba* is a white-flowering version (Figure A-109).

FIGURE A-109: *Strelitzia reginae*, Bird of Paradise

Stromanthe sanguinea 'triostar'

Pronunciation: stroe-MANTH-ee sang-GWIN-ee-a

Common name: Triostar Stromanthe

Family name: *Marantaceae*

Description: Rhizomatous, herbaceous perennials with colorful, patterned, lanced-shaped foliage.

Maintenance: Bright, indirect light; keep soil evenly moist, plant prefers high humidity.

Propagation: Division.

Pests: Spider mites, mealybugs, aphids, scale, bacterial leaf spot, soft rot, and fungal diseases.

Notes: Species is native to Central and South America (Figure A-110).

FIGURE A-110: *Stromanthe sanguinea* 'Triostar', Triostar Stromanthe

Syngonium podophyllum

Pronunciation: sin-GONE-ee-um poe-doe-FIL-lum

Common name: Nepthytis

Family name: *Araceae*

Description: Compact in juvenile stage, then trailing, arrowhead-shaped leaves, often with white variegation, on long petioles.

© Cengage Learning 2013

FIGURE A-111: *Syngonium podophyllum*, Nepthytis

Maintenance: Soilless potting mix; bright, indirect light; do not allow soil to dry completely; plant stems can be trained for upright version or can be used for hanging/cascading plant display.

Propagation: Stem tip or stem-segment cuttings.

Pests: Spider mites, mealybugs, aphids, scale, bacterial leaf spot, soft rot, and fungal diseases.

Notes: Native to Central America (Figure A-111).

Tillandsia sp.

Pronunciation: ti-LANDS-ee-a

Common name: Air Plant, Tillandsia

Family name: *Bromeliaceae*

Description: A wide genus of terrestrial, epiphytic or rock-dwelling perennial bromeliads, trichomes present on leaves, leaves generally linear, sometimes curling or curved, flowers present on colorful scapes.

Maintenance: Bright, indirect light; high humidity; may not require soil, just a stable perch.

Propagation: Remove and grow offsets, sow seed.

Pests/Disease: Scale, mealybugs, *Fusarium* rot.

Notes: Easy-to-grow plants that add interest to existing plantings such as securing small plants into aerial prop roots of *Ficus* or covering a foam sphere.

T. caput-medusae (KAP-ut ma-DO-see), Medusa's Head, with snake-like, twisted leaves.

T. cyanea (see-ANN-ee-a), Pink Quill, has curved, wire-like leaves and a flattened inflorescence consisting of bright pink bracts and blue flowers.

T. ionantha (ee-oh-NAN-tha), mostly epiphytic, with short, coarse, gray-green leaves to 1.5 inches (4 centimeters) long, violet-blue and white flowers in spring (Figure A-112).

T. usneoides (us-nee-OI-dees), Spanish Moss, Southeast United States, Central and South America.

T. xerographica (ze-roe-GRAF-i-ca), Xerographica, curling, curving, silver leaves (Figure A-113).

© Cengage Learning 2013

FIGURE A-112: *Tillandsia ionantha*, Air Plant, Tillandsia

© Cengage Learning 2013

FIGURE A-113: *Tillandsia xerographica*, Air Plant, Tillandsia

Tolmiea menziesii

Pronunciation: tol-MEE-a men-ZEEZ-ee-ee

Common name: Piggy Back Plant, Thousand Mothers

Family name: *Saxifragaceae*

Description: Herbaceous perennial where new plants are produced on leaves at blade-petiole juncture, native to Western North America where it grows in woodland forests.

Maintenance: Bright, indirect, filtered light; organic soil mix with good drainage, do not allow soil to dry completely.

Propagation: Remove and grow plantlets, pin mature leaves to light mix until new plant generates, or sow seeds.

Pests/Disease: Mealybugs, *anthracnose*.

Notes: Houseplant suited for hanging plant or tabletop.

Tradescantia pallida 'purple heart'

Pronunciation: tra-des-KANT-i-a PA-li-da

Common name: Purple Heart

Family name: *Commelinaceae*

Description: Trailing herbaceous perennial with purple stems, clasping leaves, fine hairs on both, bright pink flowers borne on terminals.

Maintenance: Bright light, well-drained soil allowing soil to dry slightly between waterings.

Propagation: Stem cuttings, division of larger plants.

Pests: Aphids, spider mites, viruses.

Notes: Synonymous with *T. p.* 'Purpurea'.

Tradescantia spathacea

Pronunciation: tra-des-KANT-i-a spa-tha-KEE-a

Common name: Moses-in-the-Cradle

Family name: *Commelinaceae*

Description: Clump-forming perennial, mostly erect, fleshy leaves, green above, purple underneath; white flowers are held in long-lasting pair of greenish purple bracts resembling a boat.

Maintenance: Bright light, well-drained soil allowing soil to dry slightly between waterings.

Propagation: Stem cuttings, division of larger plants.

Pests: Aphids, spider mites, viruses.

Notes: Native to West Indies and Central America (Figure A-114).

FIGURE A-114: *Tradescantia spathacea*, Moses-in-the-Cradle

Tradescantia zebrina

Pronunciation: tra-des-KANT-i-a ze-BREE-na

Common name: Wandering Jew

FIGURE A-115: *Tradescantia zebrina*, Wandering Jew

FIGURE A-116: *Triticum aestevum*, Wheat Grass

FIGURE A-117: *Vanda sp.*, Vanda Orchid

Family name: *Commelinaceae*

Description: Trailing perennial with ovate-oblong leaves, two longitudinal silver stripes above, royal purple undersides.

Maintenance: Bright light, well-drained soil allowing soil to dry slightly between waterings. Pinch leggy stems to encourage bushiness.

Propagation: Stem cuttings, division of larger plants.

Pests: Aphids, spider mites, viruses.

Notes: Native to southern Mexico (Figure A-115).

Triticum aestevum

Pronunciation: TRI-ti-kum EE-ste-vum

Common name: Wheat Grass

Family name: *Poaceae*

Description: Monocot grass with bright green blades.

Maintenance: Bright light, do not allow growing medium to dry out, keep planting cool to lengthen display duration.

Propagation: Seed.

Pests/Disease: Aphids, *Botrytis*.

Notes: Used for short-term displays (Figure A-116).

Vanda sp.

Pronunciation: VAN-da

Common name: Vanda Orchid

Family name: *Orchidaceae*

Description: Epiphytic, monopodial, thick stems bear semi-rigid, strap-like leaves, flowers large and showy borne on racemes, with prominently colored veins and small lips.

Maintenance: Keep plants in bright light, use an epiphytic orchid mix, water thoroughly, and allow to drain and dry slightly.

Propagation: Stem cuttings or pot offsets.

Pests/Disease: Spider mites, aphids, whiteflies, mealybugs, leaf spots, bacterial soft rot, viruses.

Notes: Found in India, Southeast Asia, Philippines, and Australia (Figure A-117).

Vriesea sp.

Pronunciation: VREEZ-ee-a

Common name: Vriesea, Bromeliad

Family name: *Bromeliaceae*

Description: Mostly epiphytic, perennial bromeliads, leaves emerge in rosettes and can be green, dark green, or variegated with green, purple, yellow, and other colors; flowers are usually borne on flattened, two-ranked, spike-like racemes, which helps to distinguish the plant from other Bromeliads.

Maintenance: Grow in epiphytic bromeliad mix, moderate light, keep central cup of foliage filled with water.

Propagation: Remove offsets and pot separately, sow seed.

Pests/Disease: Scale, mealybugs, crown rot.

Notes: Named after Willem Hendrick de Vriese (1806–1862), Dutch botany professor. Found in Mexico, Central and South America, and the West Indies (Figure A-118).

FIGURE A-118: *Vriesea sp.*, Vriesea, Bromeliad

Washingtonia sp.

Pronunciation: wa-shing-TON-ee-a

Common name: Washingtonia Palm

Family name: Palmae

Description: Single-stemmed palms; deeply lobed, fan-shaped leaves, toothed petioles.

Maintenance: Soil-based media with sharp sand for drainage; allow to dry between waterings, full light.

Propagation: Seed.

Pests/Disease: Spider mites, scale, viruses, leaf spots.

Notes: Native to arid areas of Southwest United States and northern Mexico.

Yucca sp.

Pronunciation: YOO-ka

Common name: Yucca

Family name: *Agavaceae*

Description: Rosette-forming perennials, with rigid, often sharp-pointed, lanceolate foliage.

Maintenance: Well-drained media, high light; allow soil to dry between waterings.

Propagation: Pot suckers as they arise.

Pests/Disease: Scale, fungal leaf spots.

Notes: Durable plant for high light intensities and dry heat.

Zamia furfuracea

Pronunciation: za-MEE-a fur-fur-A-key-a

Common name: Cardboard Plant

Family name: *Zamiaceae*

FIGURE A-119: *Zamia furfuracea*, Cardboard Plant

Description: Short plant bearing pinnate foliage at stem terminal; stiff, pale green leaflets with brown hairs.

Maintenance: Display in bright light in well-drained soil, allow to dry somewhat between waterings.

Propagation: Seed.

Pests/Disease: Scale, mealybugs, leaf spots.

Notes: Coastal eastern Mexico. *Z. pumila* (PEW-mi-la), Coontie, Seminole Bread is grown as a cut foliage and potted plant (Figure A-119).

Zamioculcas zamiifolia

Pronunciation: za-mee-o-KUL-cas za-me-ee-FO-lee-a

Common name: ZZ Plant

Family name: *Araceae*

Description: Herbaceous evergreen, rhizomatous, pinnately compound with shiny pinnae.

Maintenance: Tolerates low light, but prefers bright, indirect light. Water sparingly.

Propagation: Division of larger established plants or root-swollen leaf bases.

Pests/Disease: Susceptible to root rot if overwatered.

Notes: Durable interior plant; native to eastern Africa.

GLOSSARY

Abiotic: Pertaining to nonbiological conditions.

Acuminate: Tapered to a long point.

Adventitious: Plant organ in unusual position such as aerial prop roots arising from aboveground stem tissue.

Aerial root: Root arising from aboveground stem tissue.

Air layering: A method of propagation where a notch is cut in a stem, then wrapped with moistened sphagnum moss, and enclosed in plastic wrap until rooting occurs, at which time it is removed from the parent plant and grown.

Anaerobic: Absence of oxygen.

Angiosperms: Flowering plants that develop seeds in fruits.

Anion: A negatively charged molecule.

Annual: A plant completing its life cycle within one year.

Anther: The pollen-bearing appendage of the stamen.

Antheridia: Organs that produce male sex cells in ferns, mosses, and other plant life.

Archegonia: Organs that produce eggs in ferns, mosses, and other plant life.

Areole: A cluster of spines in cacti.

Armature: A framework for sculptural forms; in horticulture, wire or rigid material used for plant material support for topiary or floral design.

Asexual propagation: Production of plants by vegetative means rather than seed.

Auricle: An ear-like shape.

Azalea pot: Container that is half as tall as it is wide.

Biennial: A plant that completes its life cycle in two years, the first of which is vegetative, the second culminating in flowering, fruiting, and senescence.

Binomial nomenclature: A two-name system used to classify plants; the first word is the genus and the second word is the species within the genus.

Bioeffluents: Byproduct chemicals of human respiration.

Biophilia: Innate love, affinity for living things.

Biowall: A vertical plane of living plant material.

Bipinnate: Pinnate with divisions in pinnae; twice-pinnate.

Blade: A leaf.

Bonsai: The art of dwarfing a plant via pruning and restriction.

Bract: A leaf-like structure beneath a cluster of flowers.

Brand: Distinctive traits that identify a corporate entity.

Bulb: A short stem surrounded by fleshy leaves that acts as an underground storage structure.

Bulbil: A bulb-like organ produced in a leaf axil and a means of vegetative propagation.

Cache pot: A decorative pot that holds a plant within a grow pot.

Cactus mix: A mixture of organic matter combined with sand or other amendment to promote drainage.

Callus: Thickened tissue produced by a plant to cover a wound.

Calyx: Sepals, fused or separate, which form the perianth's outermost whorl.

Capillary action: The upward movement of water through soil or plant stems.

Cation: A positively charged molecule.

Cation exchange: The capacity of negatively charged soil particles to attract positively charged ions and to exchange one ion for another.

Caudex: Swollen base of stem in plants such as cycads and succulents.

Chloroplasts: Cellular organelles that contain chlorophyll.

Chlorosis: Loss of green chlorophyll pigment resulting in a pale appearance.

Christmasscaping: The business of displaying holiday decorations in commercial spaces.

Codominant leaders: Vigorous, upright branches nearby the central vertical axis branch; removal allows establishment of a central leader.

Column: The combined stigma and style in orchids.

Compaction: Loss of air-filled pore spaces in plant media.

Compost: Formed by the decay and breakdown of organic material.

Compound: As in a compound leaf, two or more separate leaflets borne on a single leaf stalk.

Container effect: Longer, taller soil column profiles drain faster than shorter profiles.

Corolla: The part of the flower made up of petals.

Corymb: Flat-topped or dome-shaped inflorescence of stalked flowers or flower heads arising at different levels on alternate sides of an axis.

Culm: Jointed, hollow stems as found in Bamboo.

Cultivar: Cultivated variety.

Cutting: A segment of root, shoot, or leaf used for vegetative propagation.

Cyathia: The inflorescences of Euphorbia where a cup-shaped involucre surrounds a single pistil and multiple male flowers.

Cyme: Flat-topped or dome-shaped branched inflorescence with each axis ending in a flower, the most mature in the center.

Deciduous: A plant that sheds leaves at the end of a growing season.

Desiccate: Dried-down plant tissues.

Diffusion: Molecular movement from an area of high concentration to an area of low concentration.

Dioecious: Male and female flowers are borne on separate plants.

Disc: An organ at the base of the ovary.

Entire: Nonlobed leaf.

Epiphytic: A nonparasitic plant growing on another plant.

Etiolation: Stem elongation caused by lack of light.

Evergreen: Plants that retain foliage longer than one growing season.

Exudates: Substances emitted from the cut end of a plant material stem.

Filament: A stalk that bears the anther, both of which make up the stamen.

Footbath: A container originally designed for soaking feet, often used for holding plants.

Footcandle: Light intensity measured by the illumination of a surface one foot from the source of one candela, equal to one candela per square foot.

Genus: Grouping of closely related species; the first name in a binomial.

Glabrous: Without hairs.

Glaucus: Blue-gray bloom.

Green building: Environmentally responsible and resource-efficient construction, from site-selection to ultimate demolition.

Green wall: A biowall.

Grit: Coarse media amendment, usually sand or gravel.

Gymnosperms: Plants that bear naked seeds in cones rather than in ovaries.

Herbaceous plant: Nonwoody plant that dies back at the end of the growing season. May be annual, biennial, or perennial.

Houseplant: A plant displayed in an indoor, domestic environment for decorative purposes.

Hue: A color; visible light.

Humidity: The measure of the air's moisture content, expressed as a percentage.

HVAC: Heating, ventilation, and air conditioning.

Hydrophilic gel: Highly absorbent polymers often used as soil amendments to slowly release moisture in the root zone.

Imidacloprid: a systemic insecticide which acts on the central nervous system, paralyzes and eventually kills insects. Beyond controlling insects on plants, it is also used to control fleas on domestic pets.

Indusium: The covering of the sorus in ferns.

Inflorescence: Multiple flowers on a stalk; arrangement of parts is specific to a particular variety.

Intensity: Extreme degree as in light, temperature, or brightness of color.

Internodes: Spaces between nodes, emergence points of leaves and shoots on a stem.

Involucre: Whorl of bracts beneath an inflorescence.

Jardinière: A decorative pot for plants.

Latex: A milky sap exuded from certain plants when injured, for example, *Euphobia pulcherrima*; may cause skin irritation.

Latin name: A scientific name; botanical nomenclature.

Leachate: Percolated drainage from plants.

Leaching: The process of removing ions and molecules from plant soil by heavy watering.

Leaflets: Individual blade unit of a compound leaf.

Lenticel: On bark, a pore allowing for gas exchange between the plant tissues and air.

Lip: A modified petal in orchid flowers.

Living wall: Synonymous with green wall.

Loading dock: Platform (and surrounding area) that is level with truck trailer to facilitate receiving and shipping.

Long-day plants: Plants that flower when day length exceeds a critical point.

Lyse: Cell bursting.

Mechanics: Items used to fix plant materials in place in a design.

Meristem: Tip of a root or shoot where cell division occurs.

Monocarpic: Fruiting once, and then dying.

Monopodial: A stem or a rhizome growing from an apical or terminal bud and not usually producing branches, for example, spruce trees.

Mucronate: With a short, abrupt point.

Native: Species growing naturally in a particular area.

Necrotic: Dead plant tissue.

Niche: A recess in a wall to hold a plant or art object.

Node: The swollen portion of a stem bearing buds or leaves.

Obovate: Egg-shaped in outline, broadest above the middle.

Oculus: Round opening in a wall or roof allowing for vision or light penetration.

Off-gassed: Allowing manufactured materials to emit volatile organic compounds until fully exhausted.

Offsets: Small plants that arise vegetatively from parent plants.

Orchid mix: A mixture of large, organic material such as bark for anchorage of orchid roots. Terrestrial orchid mixes may include soil or fine organic particulate while epiphytic mixes dominate with large, organic particles.

Organic: 1. Carbon-based matter. 2. Natural materials and environmentally aware practices.

Osmosis: Diffusion of water molecules across a semi-permeable membrane.

Ovate: Egg-shaped.

Palisade layer: Specialized parenchyma cells, these photosynthetic cells are just below the upper leaf tissue epidermis.

Panicle: A raceme with branches.

Pappus: A tuft of hairs or bristles on a fruit.

Peat: Derived from sedge peat or sphagnum peat, moisture-retentive, acidic, partially decayed organic matter used as media or as an amendment.

Pedicel: The stalk of a single flower in an inflorescence.

People/plant interaction: The connection between people and plants due to their mutually beneficial relationship based upon a sustainable, environmental, and social-psychological framework.

Perennial: A plant living more than two years.

Perianth: Comprised of petals and sepals, usually when the two are indistinguishable.

Perlite: Volcanic rock-derived soil amendment aiding in drainage and lessening soil compaction.

Petal: A modified leaf making up part of the corolla.

Petiole: The stalk that connects the leaf blade to the stem.

Photoperiod: The duration of light necessary to trigger a plant to flower.

Photosynthesis: The production of carbohydrates in chlorophyll pigments by the combination of carbon dioxide and water in the presence of light; oxygen is a byproduct. $6CO_2 + 6H_2O + \text{light} = C_6H_{12}O_6 + 6O_2$.

Phototropism: Plant bending toward light.

Phyllode: A flattened, leaf-like petiole.

Physic gardens: A plot of land used for the cultivation of medicinal plants in the 16th and 17th centuries.

Phytotoxicity: Harmful, destructive, or deadly to a plant.

Pilaster: A pillar or column set into a wall as an ornamental accent rather than a load-bearing support.

Pinna: Leaflet of a pinnately compound leaf such as a palm leaflet.

Pistil: The female portion of a flower.

Plant stand: A column used to display a potted plant.

Post-harvest: Time after plant or plant part has been removed from production.

Potting mix: Moisture-retaining but well-drained medium for containerized plants. Soil-based mixes contain pasteurized mineral soil and amendments such as peat, perlite, and sand; soilless mixes are usually peat-based, often with amendments.

Pot-up: To plant a newly separated, young plant into its own container.

Pre-made: A floral design created by a manufacturer and then sold to a retailer for resale to a consumer.

Propagate: To generate new plants via seed (sexual propagation) or cuttings (asexual propagation).

Prothallus: Leaf-like tissue germinating from a fern spore, which bears sexual organs and develops into a mature plant.

Pseudobulb: The thickened stem of an orchid plant.

Pups: Small, vegetatively produced plants arising from a parent bromeliad.

Raceme: Inflorescence of stalked flowers radiating from a single, unbranched axis with juvenile flowers at the tip.

Rachis: The midrib of a compound leaf as in *Nephrolepis exaltata*.

Resistance: Pests and diseases that are capable of withstanding demise from pesticides and fungicides.

Respiration: The process by which sugar (glucose) is broken down, in the presence of oxygen, to make energy. $C_6H_{12}O_6 + 6O_2 = 6CO_2 + 6H_2O + \text{energy}$.

Rhizome: Underground or ground level stem, horizontal and branching.

Root/shoot ratio: The ratio between the quantity of leaves and stems to the quantity of roots in a plant.

Rosette: Dense whorl of leaves arising from the crown of a plant as in bromeliads.

Scape: A leafless flower stalk as in *Amaryllis*.

'Scape: Plantscape.

Scientific name: The binomial; Latin name.

Senescence: Plant tissue death.

Sepal: Parts of the perianth outside of the flower petals; can be colorful like petals or green and leafy.

Sessile: Clasping leaves, attached directly at the base.

Shade: A device mounted to an interior window frame used to reduce or screen light or heat.

Short-day plants: Plants that flower when day length is less than a critical point.

Simple: A leaf having a single blade portion, without leaflets or lobes.

Site specificity: An object created for a particular place or environment.

Softened: Water containing little or no dissolved calcium or magnesium.

Sori: Plural of sorus.

Sorus: A cluster of sporangia.

Spadix: Unstalked flowers densely arranged on a fleshy stem as in *Anthurium*.

Spathe: A large bract surrounding an inflorescence as in *Spathiphyllum*.

Species: A group of closely related plants capable of interbreeding with each other, but with little success with other species; the second name in a binomial.

Specific epithet: Species name; the second name in a binomial.

Sporangia: The spore-bearing organ; plural of sporangium.

Sporophyte: The spore-producing phase in plants.

Spray: Flowers arranged on a single, branched stem.

Stamen: The male portion of a flower consisting of the anther and filament.

Standard pot: A pot approximately twice as tall as it is wide.

Stem-tip cutting: A cutting taken from new, vegetative (nonflowering) growth.

Stigma: The top-portion of the pistil, which receives pollen grains. Latin for nail.

Stolon: Aboveground stem tissue capable of producing shoots or roots.

Stomata: Epidermal pores, an opening between two guard cells; plural of stoma.

Strobilus: A cone.

Style: A part of the pistil; connects the stigma and ovary.

Subirrigation: Supplying water for plants from below.

Succulent: Plant native to arid terrain often having fleshy roots, stems, or leaves capable of water storage.

Sucker: A shoot arising from root tissue of a mature parent plant.

Sympodial: A form of growth where the terminal bud turns into a flower or dies, encouraging side branching.

Terrarium: A clear glass or plastic container in which plants are grown.

Terrestrial: A plant that grows on land, in soil.

Throat: Opening of a tubular flower part, for example, the interior of a *Cattleya* lip.

Tint: Adding white to a color to lessen its saturation.

Tone: Adding gray to a color to lessen its saturation.

Topiary: Plant material clipped or held in place to resemble geometric or figurative forms.

Transpirate: Evaporation of water through stomata.

Tropical: The region between the Tropic of Cancer and the Tropic of Capricorn characterized by high rainfall and very warm temperatures.

Tuber: Underground, swollen root or stem storage tissue.

Tubercle: Wart-like nodule.

Turgid: Swollen, distended from water uptake.

Umbel: Flat-topped or dome-shaped branched inflorescence with numerous stalked flowers terminally borne from a single point, for example, *Hoya carnosa*.

Underplanting: The presence of small plants underneath larger plants, often as groundcover.

Value: The lightness (adding white to create a tint) or darkness (adding black to make a shade) of a color.

Variegation: Irregular pigmentation, the result of mutation or disorder.

Vegetative propagation: Initiation of new plants via cuttings, division, grafting, or layering.

Vermiculite: Expanded mica; a soil amendment improving moisture retention and cation exchange capacity.

Verticalscape: A green wall.

Volatile organic compound (VOC): Compounds, both human-made and naturally-occurring, that have a low boiling point and thus enter the surrounding air from liquid or solids.

Wabi-sabi: Rustic simplicity arising from nature or irregularities in handmade objects coupled with age, wear, or repair.

Wash: A deposit of eroded debris such as rocks or driftwood.

Water tension: The result of water molecule cohesion.

Whorl: A circular arrangement of petals, leaves, shoots, or flowers.

Woody: Plants with stem tissue that differentiates and produces cork (bark).

INDEX

Note: Page numbers followed by *f* and *t* denote figures and tables, respectively.

A

D

G

N

O

P